EXTENDED-NANOFLUIDIC SYSTEMS FOR CHEMISTRY AND BIOTECHNOLOGY

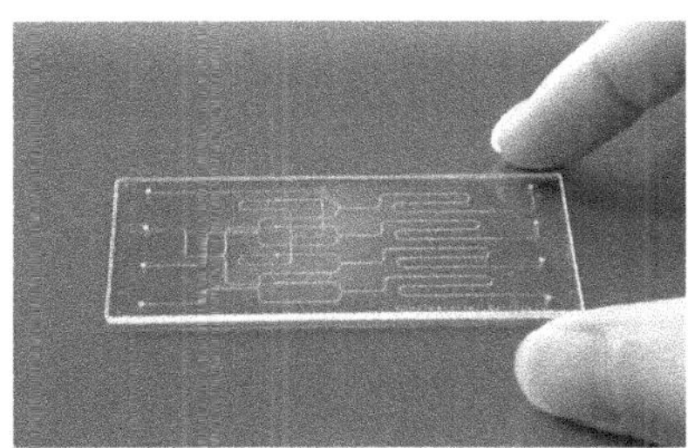

EXTENDED-NANOFLUIDIC SYSTEMS FOR CHEMISTRY AND BIOTECHNOLOGY

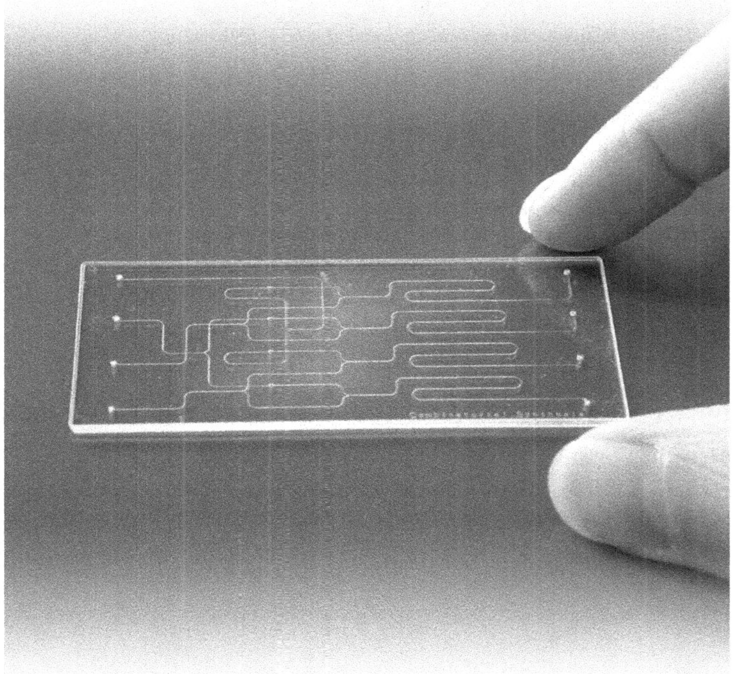

Kazuma Mawatari (The University of Tokyo, Japan)

Takehiko Tsukahara (Tokyo Institute of Technology, Japan)

Yo Tanaka (RIKEN Kobe Institute, Japan)

Yutaka Kazoe (The University of Tokyo, Japan)

Philip Dextras (The University of Tokyo, Japan)

Takehiko Kitamori (The University of Tokyo, Japan)

Imperial College Press

ICP

Published by

Imperial College Press
57 Shelton Street
Covent Garden
London WC2H 9HE

Distributed by

World Scientific Publishing Co. Pte. Ltd.
5 Toh Tuck Link, Singapore 596224
USA office: 27 Warren Street, Suite 401-402, Hackensack, NJ 07601
UK office: 57 Shelton Street, Covent Garden, London WC2H 9HE

British Library Cataloguing-in-Publication Data
A catalogue record for this book is available from the British Library.

EXTENDED-NANOFLUIDIC SYSTEMS FOR CHEMISTRY AND BIOTECHNOLOGY

ISBN-13 978-1-84816-801-5
ISBN-10 1-84816-801-2

Typeset by Stallion Press
Email: enquiries@stallionpress.com

Printed in Singapore.

CONTENTS

Chapter 1

INTRODUCTION

Integrated microchemical systems on chips (microchemical chips) are recognized as one of the key technologies for future progress in chemical and biochemical technologies. The advantages of this miniaturization include compact, high speed, and functional instrumentation for analysis and synthesis in bio and related sciences and technologies. In 1979, Terry *et al.* first reported a chemical chip in which a gas chromatography column was fabricated on a silicon wafer; there had not been any reports on microintegration for a decade at that time.[1] In the 1990s, Manz and coworkers pioneered the lab-on-a-chip, or microchip, concept and illustrated its usefulness. Manz's group integrated the function of capillary electrophoresis on a single glass chip. Along with the requirement for fast DNA analysis of a small sample with small reagent volume, the microchip has been recognized as a promising technology, mainly for the separation of DNA and proteins. The technologies utilized were primarily electroosmotic flow (EOF), electrophoretic separation, and laser-induced fluorescence (LIF) detection.[2,3] At that time, these microchips were referred to as micro total analysis systems (μTAS). However, other analytical and synthesis methods were required for wide application in more general, analytical, combinatorial, physical, and bio-related chemistries that included complicated chemical processes, organic solvents, neutral species, and non-fluorescent molecule detection. From this point of view, general microintegration methods on microchips were quite important for wide application.

For these purposes, general concepts were proposed to achieve general microintegration on a chip, which was called a lab-on-a-chip, or microchemical chip. Many bulk scale unit operations such as mixing, extraction, phase separation, and other unit operations of chemical processes are integrated as microscale chemical components and named as micro unit operations (MUOs). The MUOs can be combined in parallel and in series, like an electric circuit, through continuous flow chemical processing (CFCP). The microchemical chip also has a functional chemical central processing unit (CCPU). This combination has enabled a variety of analyses, syntheses, and biochemical systems to be integrated on microchips, and has been proven to be an effective general methodology for microintegration (Figure 1.1). As a result, superior performance has been demonstrated in a shorter processing time (from days or hours to minutes or seconds), smaller sample or reagent volume (diagnosis with one drop of blood), with easier operation (from professional to personal), and in smaller systems (from 10 m scale chemical plants to desktop plants, and desktop systems to mobile systems) than conventional analysis, diagnosis, and chemical synthesis systems. Practical prototype systems have also been realized in environmental analysis, clinical diagnosis, cell analysis, gas analysis, medicine synthesis, microparticle synthesis, and so on. The general concept has allowed the establishment of "chemical devices" for the first time.

In order to realize these basic concepts, fundamental technologies are essential. These technologies include: pressure-driven microfluidics

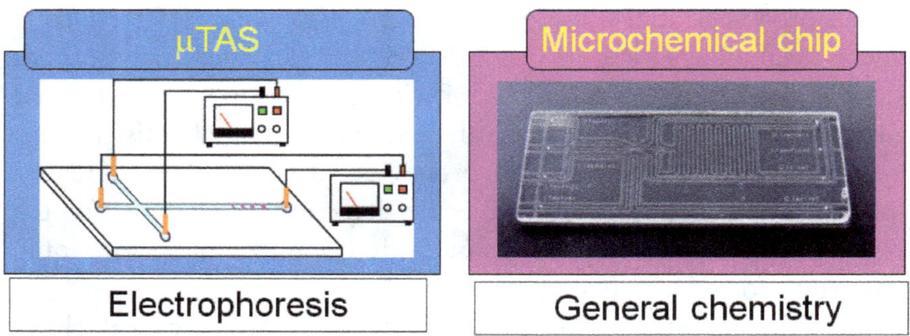

Figure 1.1. Transition from electrophoresis to general chemistry by microchemical chip or lab-on-a-chip.

(parallel multiphase flow or droplet-based multiphase flow), surface control methods (hydrophobic/hydrophilic, bio-molecule, cell, catalyst, etc.), detection methods (optical, electrochemical, and conventional analytical instruments by developing the interface), and fabrication methods (silicon, glass, polymer, ceramics, etc.). The device development of these methods is also an important issue because microchemical chips work as CCPUs and the peripheral devices are necessary to realize microchemical systems.

There are two directions for microtechnologies. One is to put these technologies to practical and commercial use for micrometer scale chemical systems on chips. For example, practical systems have been developed for clinical diagnosis, environmental analysis, food analysis, drug synthesis, basic research for biology, and pharmaceutical and tissue engineering. For these purposes, designing tools for microchemical processes and reliable fluidic devices will be important, in addition to reducing costs.[4]

The other direction is to extend the method to nanometer scale chemical experiments, which is opening new horizons for chemical research tools. Recently, microfluidic systems and detection devices were applied to 10^1–10^3 nm scale fluidic systems, which we call extended nanospace to distinguish it from the 10^0–10^1 nm scale space belonging to conventional nanotechnology. This extended nanospace bridges the gap between single molecules and normal, condensed phases (Figure 1.2), the liquid properties of which have not yet been properly explored. In order to understand these liquid properties, new fundamental technologies are required as basic research tools, since conventional technologies are difficult to apply due to the extremely small size of the extended nanospace. These technologies are those of fabrication, fluidic control, detection, and surface modification methods, with many challenges present due to the small and closed space. Great efforts have been made to develop these basic technologies in recent years, and the methodologies have revealed many unique liquid properties.[5,6] By utilizing these unique liquid properties, new chemical operations are increasingly reported, which are quite difficult to achieve using microtechnologies, and will be useful for future bio and analytical technologies

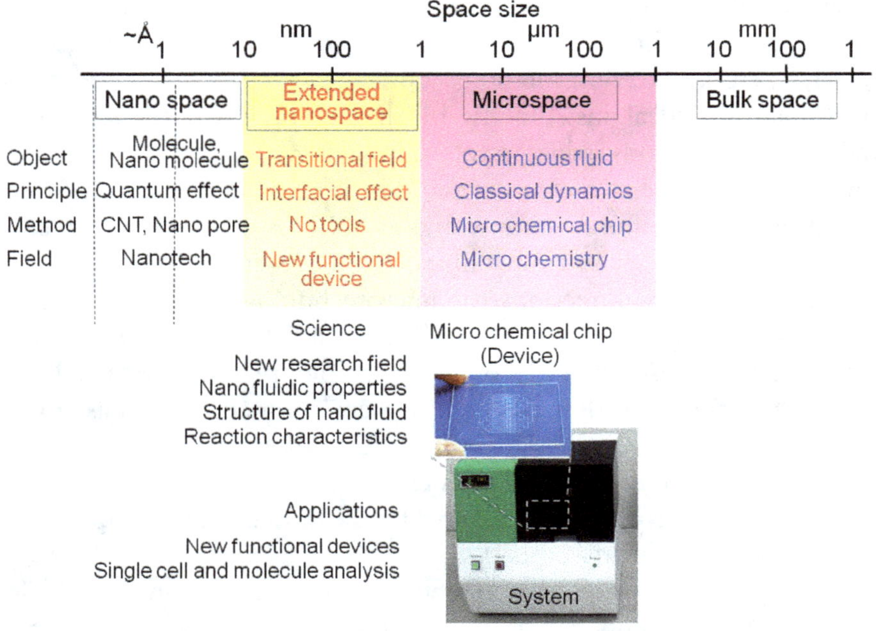

Figure 1.2. Size hierarchy and micro and extended nanospace.

(e.g. single cell and single molecule analysis). Now, microchip technologies are moving to the next generation by combining with extended nanospace.

References

1. Terry S.C., Jerman J.H., and Angell J.B. (1979), A complete GC system on a silicon wafer with a glass cover plate, including column, injector, and detector, *IEEE T Electron Dev*, **ED-26**, 1880–1886.
2. Reyes D.R., Lossifidis D., Auroux P.A., and Manz A. (2002), Micro total analysis systems. 1. Introduction, theory, and technology, *Anal Chem*, **74**, 2623–2636.
3. Auroux P.A., Lossifidis D., Reyes D.R., and Manz A. (2002), Micro total analysis systems. 2. Analytical standard operations and applications, *Anal Chem*, **74**, 2637–2652.

4. Ohashi T., Mawatari K., Sato K., Tokeshi M., and Kitamori T. (2009), A micro-ELISA system for the rapid and sensitive measurement of total and specific immunoglobulin E, and clinical application to allergy diagnosis, *Lab Chip*, **9**, 991–995.
5. Tsukahara T., Mawatari K., and Kitamori T. (2010), Integrated extended-nano chemical systems on a chip, *Chem Soc Rev*, **39**, 1000–1013.
6. Mawatari K., Tsukahara T., Sugii Y., and Kitamori T. (2010), Extended-nano fluidic systems for analytical and chemical technologies, *Nanoscale*, **2**, 1588–1595.

Chapter 2
MICROCHEMICAL SYSTEMS

Integrated microchemical systems have great potential for application in various fields. In order to realize the general analytical, combinatorial, physical, and biochemical applications, general integration methods on microchips are important. Conventional, macroscale chemical plants or analytical systems are constructed by combining unit operations, such as mixers, reactors, and separators. A similar methodology can be applied to microchemical systems. However, miniaturizing conventional unit operations is often ineffective and sometimes impossible, due to the many physical properties (e.g. heat and mass transfer efficiency, specific interfacial area, and gravitational force) and since the dominant factors for fluidics and chemistry are significantly different in microspace. Therefore, new MUOs taking these issues into account are required.

For this purpose, multiphase microflow is utilized. In conventional macroscale devices, the aqueous and organic phases are separated by gravity. In microspace, however, the fluid is greatly influenced by liquid/solid, liquid/gas, and liquid/liquid interfaces because of the large specific interfacial area. The main physical forces in the microchannels, including the viscous force, and the interfacial parameters can be analyzed using the dimensionless Reynolds (*Re*) and Bond (*Bo*) numbers, defined as the ratio of inertial-to-viscous forces and the ratio of gravity-to-tension, respectively. *Bo* is defined as $Bo = (\Delta\rho)gdh2/\gamma$, where $\Delta\rho$, γ, and dh are the density difference, the interfacial tension between the two phases, and the equivalent diameter, respectively, and where g is the gravitational acceleration (9.8 ms^{-2}). Therefore, multiphase microflows are generally considered to be laminar flows. Usually, multiphase microflows are divided into droplet flows and parallel flows, which have both advantages and disadvantages,

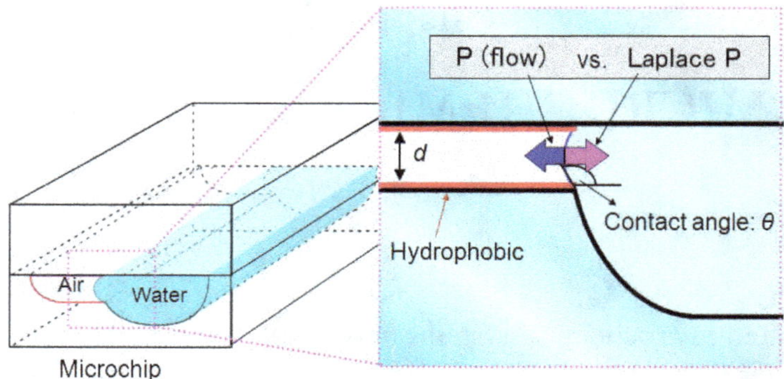

Figure 2.1. Laplace pressure for stabilizing the interface.

while parallel flows are used for the general integration of complicated chemical processes, controlled by pressure.[1] It is important to remember that the pressure driving the fluids decreases in the downstream part of the flow due to the fluids' viscosity. When two fluids in contact with one another have different viscosities, the pressure difference (ΔP_{Flow}) between the two phases is a function of the contact length and the flow velocity. Another important parameter is the Laplace pressure ($\Delta P_{\text{Laplace}}$) caused by the interfacial tension between two phases. The interface is fixed at a position in the microchannel determined by the balance established between $\Delta P_{\text{Laplace}}$ and ΔP_{Flow}. Figure 2.1 illustrates the pressure balance at the liquid/liquid interface of a two-phase microflow. With a glass surface, the liquid/liquid interface curves toward the organic phase because of the hydrophilicity of the glass, a substrate used for general chemistry. $\Delta P_{\text{Laplace}}$ is generated at the curved liquid/liquid interface. On the basis of the Young–Laplace equation, $\Delta P_{\text{Laplace}}$ is estimated as follows:

$$\Delta P_{\text{Laplace}} = \frac{\gamma}{R} = \frac{2\gamma \sin(\theta - 90^\circ)}{d}, \tag{1}$$

where R is the radius of curvature of the liquid/liquid interface, and θ is the contact angle. The contact angle is restricted to the set of values between the advancing contact angle of the aqueous phase, θ_{aq}, and that of the organic phase, θ_{org}. Therefore, $\Delta P_{Laplace}$ is restricted as follows:

$$\frac{2\gamma\sin(\theta_{aq}-90^\circ)}{d} < \Delta P_{Laplace} < \frac{2\gamma\sin(\theta_{org}-90^\circ)}{d}. \quad (2)$$

When ΔP_{Flow} exceeds the maximum value of $\Delta P_{Laplace}$, the organic phase flows toward the aqueous phase. When ΔP_{Flow} is lower than the minimum value of $\Delta P_{Laplace}$, the aqueous phase flows toward the organic phase. When the ratio of flow rates is changed, the pressure balance is maintained through changing the position of the liquid/liquid interface. This model indicates that the important parameters for microfluid control are the interfacial tension, the dynamic contact angle, and the depth of the microchannel. This model can also be applied to gas/liquid microflows. In order to stabilize the interface, $\Delta P_{Laplace}$ should be maximized by controlling channel size, shape, and surface hydrophobicity. This strategy allows various multiphase parallel flows.

Using these methods, bulk scale unit operations can be integrated as MUOs, as shown in Figure 2.2. By combining MUOs with different functions in series and in parallel, various chemical processes can be integrated into microchips through the use of a multiphase microflow network, CFCP, as shown in Figure 2.3.[2] Many MUOs have been developed for wide application, such as mixing and reaction,[3,4] phase confluence and separation,[5–16] solvent extraction,[17] gas/liquid extraction,[18–21] solid-phase extraction and reaction on surfaces,[22–33] heating,[34–38] and cell culture.[39–43] These methodologies allow for the general integration of chemical processes. Finally, the microchip is installed in a microsystem and works as a CCPU. Several microsystems have been realized for practical applications (Figure 2.4).

Designing microchips is an important issue and an example of designing a microcobalt, wet analysis is illustrated in Figure 2.5. Conventional analytical procedures in bulk scale analysis consist of a chelating reaction, solvent extraction of the complex, and the decomposition and removal of the co-existing metal complex. These

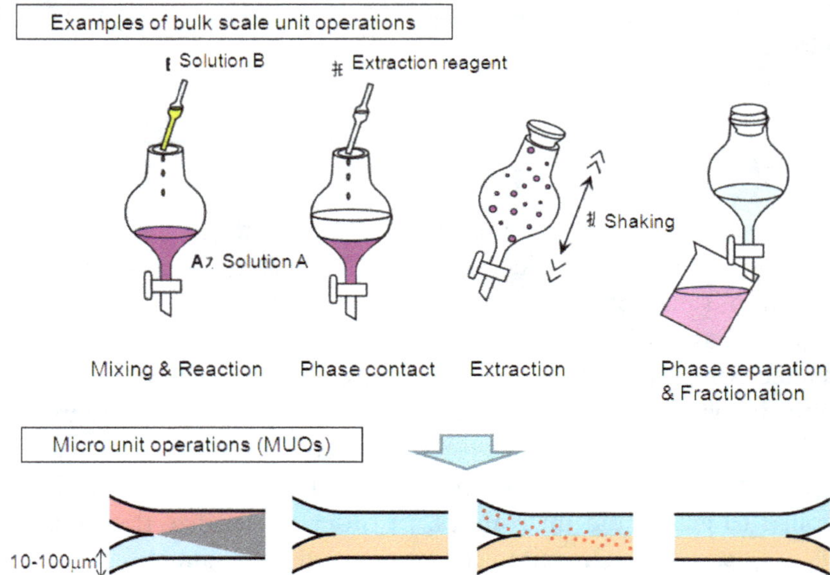

Figure 2.2. Conversion of bulk scale and MUOs.

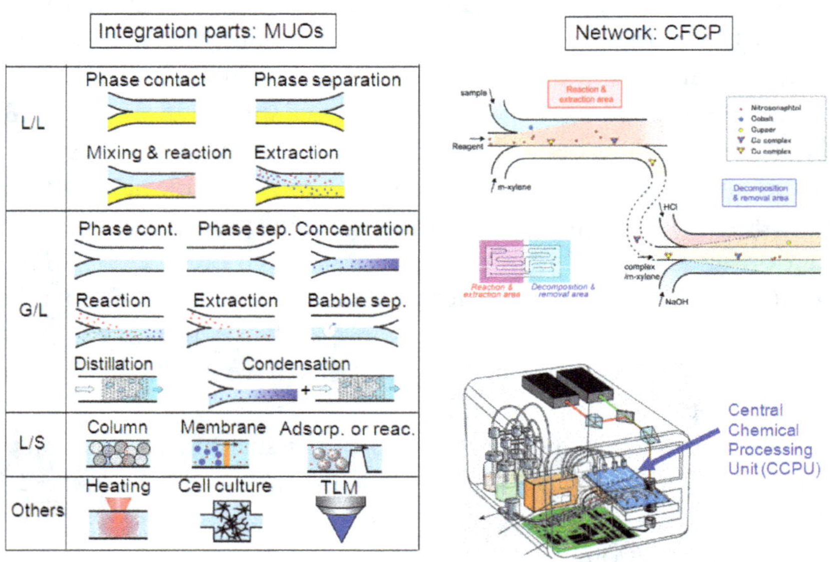

Figure 2.3. Realization of chemical processes on a microchip and in a microsystem.

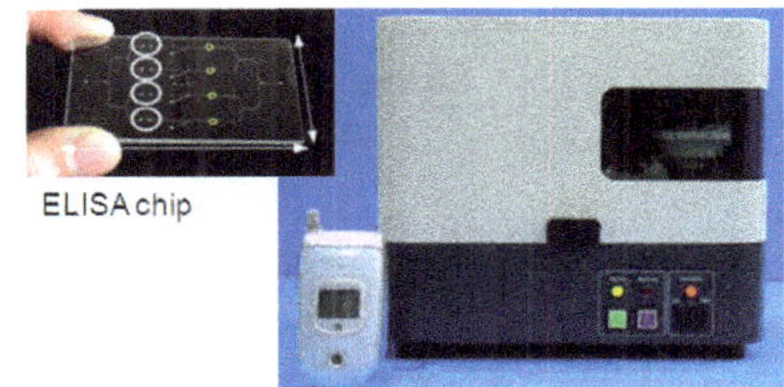

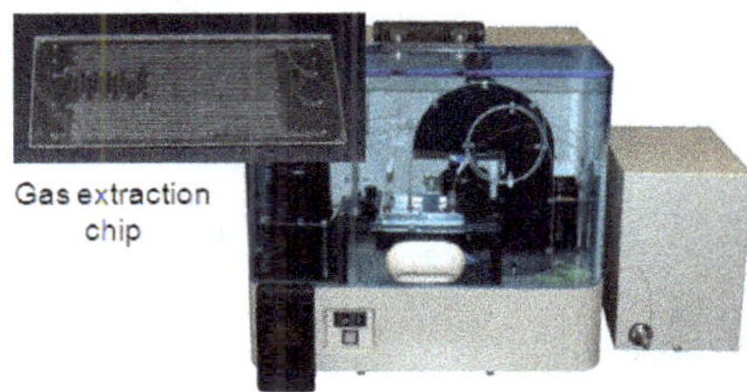

Figure 2.4. Examples of microchips and systems for real application.

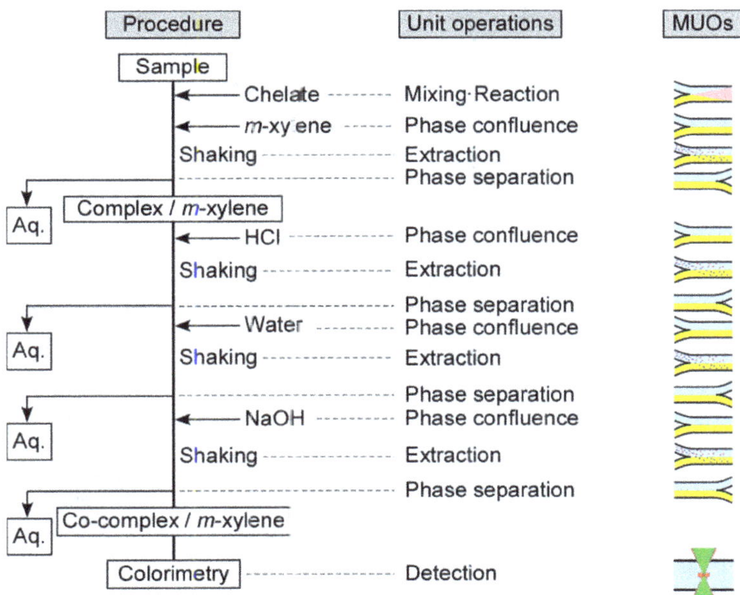

Figure 2.5. Examples of the design procedure for a micro wet analysis chip.

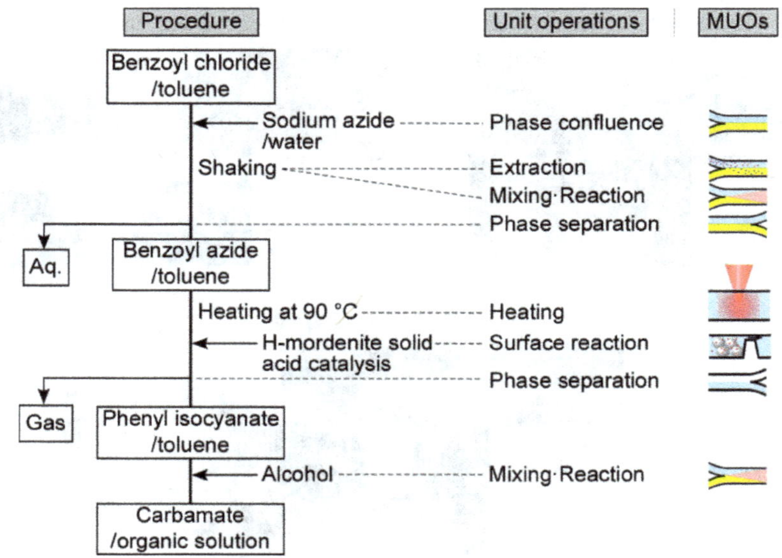

Figure 2.6. Examples of the design procedure for a carbamate synthesis chip.

procedures can be separated into unit operations: mixing and reaction, phase confluence, solvent extraction, and phase separation. These unit operations can then be transformed into MUOs. By connecting the MUOs in serial or parallel by CFCP, a microchip is designed. For blood diagnosis, multistep synthesis of carbamate is another example of CFCP, which has been demonstrated (Figure 2.6).[44] Conventional procedures for such synthesis consist of a phase transfer reaction between aqueous azide and acid chloride, a reaction on surfaces for the formation of aryl isocyanates, and the reaction between aryl isocyante and alcohol. These procedures are also examples of conventional unit operations: phase confluence, mixing and reaction, phase separation, decomposition by heating, and reaction on surfaces.

The CFCP approach can be applied to develop more complicated processing systems. More rapid analysis, bioassays, and immunoassays as well as more efficient reaction and extraction, can be achieved through CFCP systems compared to conventional devices. At present, many fundamental technologies and devices have been developed to realize reliable and high performance systems, and several companies have commercialized microsystems.

References

1. Aota A., Mawatari K., Takahashi S., Matsumoto T., Kanda K., Anraku R., Hibara A., Tokeshi M., and Kitamori T. (2009), Phase separation of gas-liquid and liquid-liquid microflows in microchips, *Microchim Acta*, **164**, 249–255.
2. Tokeshi M., Minagawa T., Uchiyama K., Hibara A., Sato K., Hisamoto H., and Kitamori T. (2002), Continuous-flow chemical processing on a microchip by combining microunit operations and a multiphase-flow network, *Anal Chem*, **74**, 1565–1571.
3. Sato K., Tokeshi M., Kitamori T., and Sawada T. (1999), Integration of flow injection analysis and zeptomole-level detection of the Fe(II)-o-phenanthroline complex, *Anal Sci*, **15**, 641–645.
4. Sorouraddin H.M., Hibara A., Proskrunin M.A., and Kitamori T. (2000), Integrated FIA for the detection of ascorbic acid and dehydroascorbic acid in microfabricated glass channel by thermal-lens microscopy, *Anal Sci*, **16**, 1033–1037.
5. Hisamoto H., Horiuchi T., Uchiyama K., Tokeshi M., Hibara A., and Kitamori T. (2001), On-chip integration of sequential ion sensing system based on intermittent reagent pumping and formation of two-layer flow, *Anal Chem*, **73**, 5551–5556.
6. Hibara A., Nonaka M., Hisamoto H., Uchiyama K., Kikutani Y., Tokeshi M., and Kitamori T. (2002), Stabilization of liquid interface and control of two-phase confluence and separation in glass microchips by utilizing octadecylsilane modification of microchannels, *Anal Chem*, **74**, 1724–1728.
7. Tokeshi M., Minagawa T., and Kitamori T. (2000), Integration of a microextraction system on a glass chip: ion-pair solvent extraction on Fe(II) with 4,7-diphenyl-1,10-phenanthrolinedisulfonic acid and tri-*n*-octylmethylammonium chloride, *Anal Chem*, **72**, 1711–1714.
8. Tokeshi M., Minagawa T., and Kitamori T. (2000), Integration of a microextraction system: solvent extraction of Co-2-nitroso-5-dimethylaminophenol complex on a microchip, *J Chromatogr A*, **894**, 19–23.

9. Surmeian M., Slyadnev M.N., Hisamoto H., Hibara A., Uchiyama K., and Kitamori T. (2002), Three-layer flow membrane system on a microchip for investigation of molecular transport, *Anal Chem*, **74**, 2014–2020.
10. Aota A., Nonaka M., Hibara A., and Kitamori T. (2007), Countercurrent laminar microflow for highly efficient solvent extraction, *Angew Chem Int Ed*, **46**, 878–880.
11. Miyaguchi H., Tokeshi M., Kikutani Y., Hibara A., Inoue H., and Kitamori T. (2006), Microchip-based liquid-liquid extraction for gas-chromatography analysis of amphetamine-type stimulants in urine, *J Chromatogr A*, **1129**, 105–110.
12. Hotokezaka H., Tokeshi M., Harada M., Kitamori T., and Ikeda Y. (2005), Development of the innovative nuclide separation system for high-level radioactive waste using microchannel chip-extraction behavior of metal ions from aqueous phase to organic phase in microchannel, *Prog Nucl Energy*, **47**, 439–447.
13. Aota A., Hibara A., and Kitamori T. (2007), Pressure balance at the liquid-liquid interface of micro counter-current flows in microchips, *Anal Chem*, **79**, 3919–3924.
14. Hibara A., Nonaka M., Tokeshi M., and Kitamori T. (2003), Spectroscopic analysis of liquid/liquid interfaces in multiphase microflows, *J Am Chem Soc*, **125**, 14954–14955.
15. Maruyama T., Uchida J., Ohkawa T., Futami T., and Katayama K. (2003), Enzymatic degradation of *p*-chlorophenol in a two-phase flow microchannel system, *Lab Chip*, **3**, 308–312.
16. Kralj J.G., Sahoo H.R., and Jensen K.F. (2007), Integrated continuous microfluidic liquid-liquid extraction, *Lab Chip*, 7, 256–263.
17. Hibara A., Tokeshi M., Uchiyama K., Hisamoto H., and Kitamori T. (2001), Integrated multilayer flow system on a microchip, *Anal Sci*, **17**, 89–93.
18. Hachiya H., Matsumoto T., Kanda K., Tokeshi M., and Yoshida Y. (2004), Micro environmental gas analysis system by using gas-liquid two phase flow, *Proc microTAS 2004*, 99–101.
19. Aota A., Mawatari K., Kihira Y., Sasaki M., and Kitamori T. (2010), Micro continuous gas analysis system of ammonia in cleanroom, *Proc microTAS 2010*, 609–611.

20. Ohira S.I. and Toda K. (2005), Micro gas analysis system for measurement of atmospheric hydrogen sulfide and sulfur dioxide, *Lab Chip*, **5**, 1374–1379.
21. Timmer B., Olthuis W., and van den Berg A. (2004), Sampling small volumes of ambient ammonia using a miniaturized gas sampler, *Lab Chip*, **4**, 252–255.
22. Sato K., Tokeshi M., Odake T., Kimura H., Ooi T., Nakao M., and Kitamori T. (2000), Integration of an immunosorbent assay system: analysis of secretory human immunoglobulin A on polystyrene beads in a microchip, *Anal Chem*, **72**, 1144–1147.
23. Sato K., Tokeshi M., Kimura H., and Kitamori T. (2001), Determination of carcinoembryonic antigen in human sera by integrated bead-bed immunoassay in a microchip for cancer diagnosis, *Anal Chem*, **73**, 1213–1218.
24. Sato K., Yamanaka M., Takahashi H., Tokeshi M., Kimura H., and Kitamori T. (2002), Microchip-based immunoassay system with branching multichannels for simultaneous determination of interferon γ, *Electrophoresis*, **23**, 734–739.
25. Sato K., Yamanaka M., Hagino T., Tokeshi M., Kimura H., and Kitamori T. (2004), Microchip-based enzyme-linked immunosorbent assay (microELISA) system with thermal lens detection, *Lab Chip*, **4**, 570–575.
26. Kakuta M., Takahashi H., Kazuno S., Murayama K., and Ueno T. (2006), Development of the microchip-based repeatable immunoassay system for clinical diagnosis, *Meas Sci Technol*, **17**, 3189–3194.
27. Ohashi T., Mawatari K., Sato K., Tokeshi M., and Kitamori T. (2009), A micro-ELISA system for the rapid and sensitive measurement of total and specific immunoglobulin E and clinical application to allergy diagnosis, *Lab Chip*, **9**, 991–995.
28. Haes A.J., Terray A., and Collins G.E. (2006), Bead-assisted displacement immunoassay for staphylococcal enterotoxin on a microchip, *Anal Chem*, **78**, 8412–8420.
29. Murakami Y., Endo T., Yamamura N., Nagatani N., Takamura Y., and Tamiya E. (2004), On-chip micro-flow polystyrene bead-based immunoassay for quantitative detection of tacrolimus (FK506), *Anal Biochem*, **334**, 111–116.

30. Sung K. and Park J.K. (2005), Magnetic force-based multiplexed immunoassay using superparamagnetic nanoparticles in microfluidic channel, *Lab Chip*, **5**, 657–664.
31. Legendre L.A., Bienvenue J.M., Roper M.G., Ferrance J.P., and Landers J.P. (2006), A simple, valveless microfluidic sample preparation device for extraction and amplification of DNA from nanoliter-volume samples, *Anal Chem*, **78**, 1444–1451.
32. Tan A., Benetton S., and Henion J.D. (2003), Chip-based solid-phase extraction pretreatment for direct electrospray mass spectrometry analysis using an array on monolithic columns in a polymeric substrate, *Anal Chem*, **75**, 5504–5511.
33. Oleschuk R.D., Shultz-Lockyear L.L., Ning Y., and Harrison D.J. (1999), Trapping of bead-based reagents within microfluidic systems: on-chip solid-phase extraction and electrochromatography, *Anal Chem*, **72**, 585–590.
34. Tanaka Y., Slyadnev M.N., Hibara A., Tokeshi M., and Kitamori T. (2000), Non-contact photothermal control of enzyme reaction on a microchip by using a compact diode laser, *J Chromatogr A*, **894**, 45–51.
35. Slyadnev M.N., Tanaka Y., Tokeshi M., and Kitamori T. (2001), Photothermal temperature control of a chemical reaction on a microchip using an infrared diode laser, *Anal Chem*, **73**, 4037–4044.
36. Goto M., Sato K., Murakami A., Tokeshi M., and Kitamori T. (2005), Development of a microchip-based bioassay system using cultured cells, *Anal Chem*, **77**, 2125–2131.
37. Burns M.A., Johnson B.N., Brahmasandra S.N., Handique K., and Webster J.R. (1998), An integrated nanoliter DNA analysis device, *Science*, **282**, 484–487.
38. Yoon D.S., Lee Y.S., Lee Y., Cho H.J., and Sung S.W. (2002), Precise temperature control and rapid thermal cycling in a micromachined DNA polymerase chain reaction chip, *J Micromech Microeng*, **12**, 813–823.
39. Tanaka Y., Sato K., Yamato M., Okano T., and Kitamori T. (2004), Drug response assay system in a microchip using human hepatoma cells, *Anal Sci*, **20**, 411–423.
40. Tanaka Y., Sato K., Yamato M., Okano T., and Kitamori T. (2006), Cell culture and life support system for microbio reactor and bioassay, *J Chromatogr A*, **1111**, 233–237.

41. Goto M., Tsukahara T., Sato K., and Kitamori T. (2008), Micro- and nanometer scale patterned surface in a microchannel for cell culture in microfluidic devices, *Anal Bioanal Chem,* **390**, 817–823.
42. Breslauer D.N., Lee P.J., and Lee L.P. (2006), Microfluidics-based systems biology, *Mol Biosyst,* **2**, 97–112.
43. El-Ali J., Sorger P.K., and Jensen K.F. (2006), Cells on chips, *Nature,* **442**, 403–411.
44. Sahoo H.R., Kralj J.G., and Jensen K.F. (2007), Multistep continuous-flow microchemical synthesis involving multiple reactions and separations, *Angew Chem Int Ed*, **46**, 5704–5708.

Chapter 3

FUNDAMENTAL TECHNOLOGY: NANOFABRICATION METHODS

3.1. Top-Down Fabrication

3.1.1. *Introduction*

A variety of fabrication methods have been investigated for achieving well-defined nanostructures on a substrate. They are broadly classified into two types of top-down and bottom-up processes.[1,2] Top-down processes are based on the nanostructures and are directly fabricated on a bulk material using lithography, etching, and/or direct milling, while, in the case of bottom-up processes, the nanostructures are designed and built up on a material by manipulating and self-assembling atoms and molecules. In order to establish electronic-photonic devices with integrated circuits, such as IC chips, the nanostructures are fabricated on a substrate as an open system, with the fabrication resolution one of the most important factors. On the other hand, since solutions and fluids must be controlled in chemical and biochemical analysis devices, closed nanospaces with functionality and operability play a more important role compared to fabrication resolution. Accordingly, the development of nano-in-micro structures, where well-ordered, size hierarchical structures and/or channels from extended-nano level to micro level are integrated on a chip, appears to be an effective strategy.

The focus has been on fabrication methods for one-dimensional (1-D) nano-in-micro structures, those with at least one dimension in the nanometer scale, depth, and two-dimensional (2-D) structures, those having both width and depth dimensions in the nanometer scale. These have been accomplished by means of various top-down approaches,

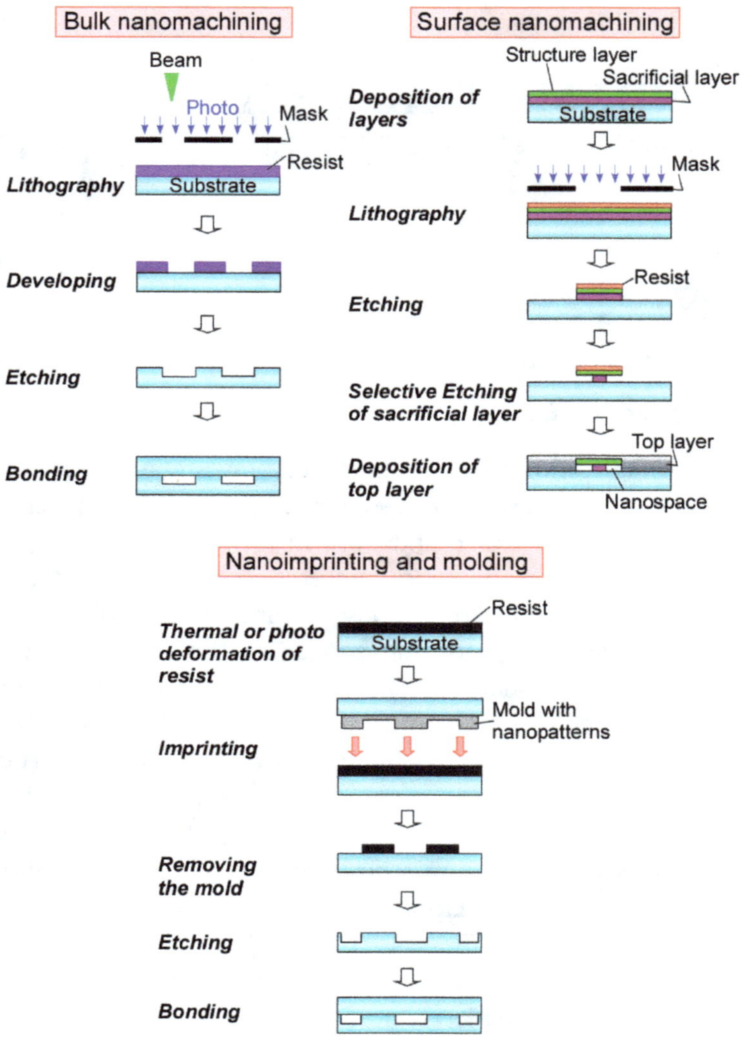

Figure 3.1. Schematic illustrations of top-down nanofabrication procedures including bulk nanomachining, surface nanomachining, and nanoimprinting and molding.

including (1) bulk nanomachining, (2) surface nanomachining, and (3) nanoimprinting and molding, as illustrated in Figure 3.1. Bulk nanomachining is the method that can be used for fabrication of nanochannels either by combined lithography-etching processes, or by direct processes. When lithographic methods are used, resist spin coating, lithography, etching, and bonding are the relevant procedures. Surface nanomachining techniques allow an enclosed nanochannel to be built by deposition, patterning, and etching the sacrificial materials on a single substrate. In nanoimprinting and molding techniques, the topographic nanopatterns are transferred from a rigid mold to a soft material, such as thermoplastic polymer film, through pressurization. The nanochannels can easily be produced in large quantities through repeated utilization of the molds. Recently, some researchers have proposed new strategies for the fabrication of nanofluidic chips, such as non-lithographic methods and hybrid-material methods. In this section, we will describe the basic techniques and practices used for nano-in-micro structural fabrication.

3.1.2. *Bulk nanomachining techniques*

3.1.2.1. *Combination of lithography and wet etching*

Photolithography, in which a substrate, coated with a photosensitive polymer resist, is exposed to ultraviolet (UV) light (*g*-line: 436 nm, *i*-line: 365 nm) through a photomask, has been recognized as a standard patterning technique. This method is widely used for microfabrication because of its numerous desirable features, such as being inexpensive, simple, and fast. Recently, many researchers have evolved to 1-D extended nanospace channel patterning, with 10–100 nm scale depth, using the same procedures as given in microchannel patterning.

Since the patterning resolution is principally determined by the combination of resist sensitivity against exposure time, and resist development contrast against developer solution, various kinds of resist can be selected according to the patterns, sizes, and materials used. There are two resist types: positive and negative. Since positive-type resists make it possible to produce small geometrical features, they are now the dominant resist used in nanolithographic processes. The positive-type resists exposed through lithographic methods cause a change in the chemical structure, leading them to be removed in the developer solution, exposing the underlying surface material. Negative-type resists show the opposite behavior compared with positive-type resists; that is, they are polymerized by lithographic exposure and become increasingly insoluble in the developer. The developer leaves the negative-type resist on the surface, wherever it is exposed, and removes only the unexposed portions. Masks used for negative-type resists need to be designed so that they represent the inverse of the desired pattern.

Typical photosensitive polymer resists (photoresists) such as OFPR-800 (Tokyo Ohka Kogyo Co., Ltd.), a positive resist, and SU-8 (MicroChem Corp.), a negative resist, are also utilized for 1-D nanochannel patterning. After the bulk materials are chemically cleaned to remove any particulate matter on the surface, as well as any traces of organic, ionic, and metallic impurities, the photoresists are spin coated uniformly onto a bulk material using a spin coater. The spin coated material is then pre-baked on a hotplate at various temperatures (typically ~100 °C) for adhering the resists to the surface, drying the resists from the bulk material, and removing the solvent. Resist thickness is controlled by the resist viscosity, solvent concentration, rotation speed of the spin coater, and so on. After the photoresists are spin coated onto a substrate, the photomask, a plate designed with microscopic patterns, is accurately aligned against the substrate. The resist-coated substrate is exposed to high intensity UV, through the photomask, and the fine microscopic patterns are transferred to the resist. The resist is developed and rinsed after the exposure. Tetramethylammonium hydroxide (TMAH) and o-xylene are used as developer solutions for chemically-amplified positive resists and non-chemically-amplified positive resists, respectively, with water and isopropyl alcohol adopted as rinse solutions. It is also necessary to control the exposure times and temperatures to an accuracy of ±10 sec and ±0.1 °C. Finally, the patterned substrates are dried by a gas blower, and etched by either wet or dry etching processes.

Specifically, the fabrication of 1-D extended nanospace channels on a Si wafer was accomplished by photolithography and wet etching processes, containing SiO_2-mask etching in hydrogen fluoride (HF) solution, and Si wafer etching in TMAH solution.[3,4] Isotropic wet etching, where the underlying bulk material surface with micropatterns is dissolved in etching solutions, such as HF and nitric acid, proceeded at the same etching rate in all directions on the material, with the shape forming a quarter-circle. When the bulk materials were immersed in etching solutions, such as TMAH and potassium hydroxide (KOH), anisotropic etching, where the etching rates differed noticeably in each direction on the material, resulted in rectangular-shaped structures. The anisotropic etching rate of the 1-D nanochannels was determined as a few nm min^{-1}, with the channel depth controlled up to 500 nm to an accuracy of a few per cent. The etched wafer fusion bonded with the Si cover wafer, or borofloat cover glass, to form the enclosed 1-D nanochannels.

Other solutions, including ammonium fluoride (NH_4F), nitric acid (HNO_3), and buffered oxide etchant (BOE), can also be used to wet etch glass substrates (Pyrex or Borofloat) at a rate of a few nm min^{-1}.[5,6] The wet-etched glass, with straight-shaped 1-D extended nanospace channels, and cover glass, with U-shaped microchannels, were polished and washed. They were then bonded at a temperature range from ambient to 550 °C. In this process, a relatively low aspect ratio of 0.001 (20 nm depth to 20 μm width) could be achieved. Detailed information concerning photolithography-based fabrications is available from the large number of technical books on microfluidics and microfabrication.

3.1.2.2. *Combination of lithography and dry etching*

As an alternative to wet etching, 1-D extended nanostructures have also been made using a couple of photolithography and dry etching techniques, such as reactive ion etching (RIE) and fast atom beam (FAB). These dry etching processes allow anisotropic etching on a substrate. In particular, RIE in a parallel-plate reactor is one of the most widely used methods. Figure 3.2(a) shows schematic illustrations of RIE. Several fluorinated gases, such as CF_4 and C_4F_8, are introduced into the reactor at 1–100 Pa under vacuum conditions. These form a plasma between the anode and cathode plates due to an applied, high frequency electromagnetic field. Since the gas molecules

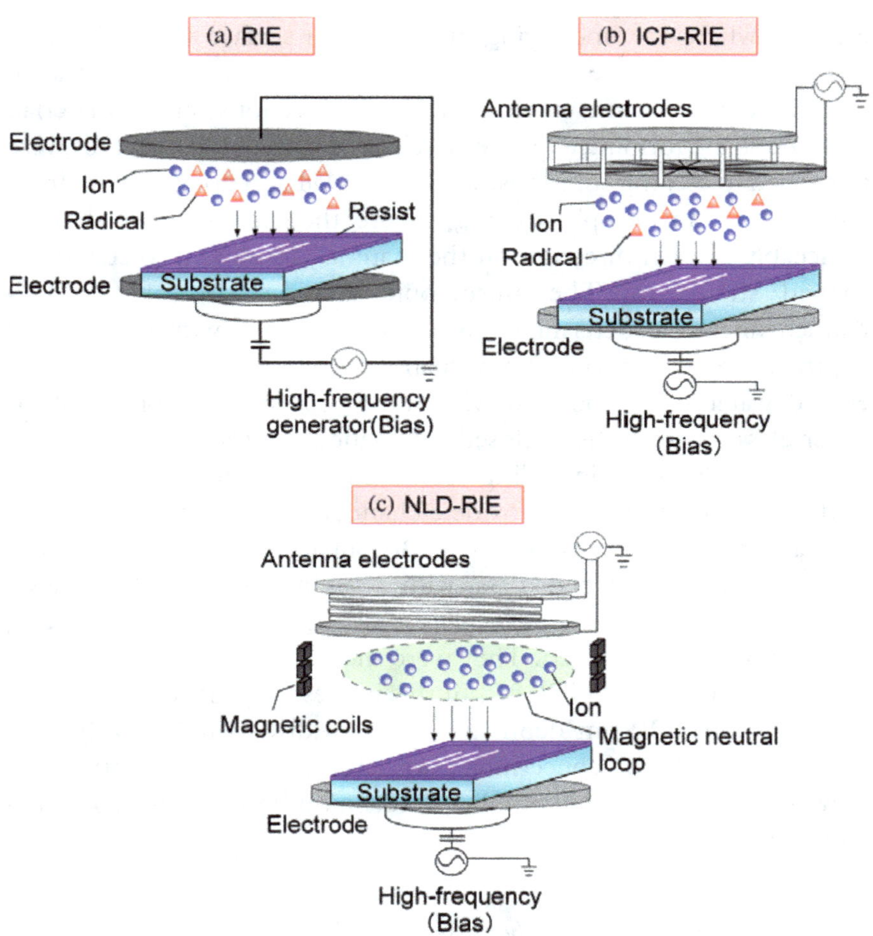

Figure 3.2. Schematic illustrations of basic principles for (a) RIE, (b) ICP-RIE, and (c) NLD-RIE dry etching techniques.

are broken down into fluorine ions and/or radicals, they are accelerated toward the anode plate and collide with the substrate. At this time, the substrate will be anisotropically etched by both a physical etching effect due to ions, and a chemical etching effect due to radicals. However, the chemical effect causes certain problems, which damage the sidewall in the etched pattern and increase the surface roughness on the etched pattern to high levels. Special classes of RIE, such as ICP (inductively coupled plasma)-RIE and NLD (magnetic neutral loop discharge)-RIE, have recently been applied to realize the etching with low damage, low surface roughness, and high aspect ratios. These special classes of RIE can reduce the production of radicals that results from chemical etching. As shown in Figure 3.2(b), the ICP-RIE system is composed of two power sources, bias electrodes and antenna electrodes. By controlling these electrodes independently, high density and high ion energy plasma can be produced at a low pressure. The NLD system is based on the plasma generated along a magnetic neutral loop by an applied radio frequency (RF) electric field (Figure 3.2(c)). Since the NLD has the important advantage that magnetic field configurations are spatially controlled, high-density plasmas can be produced with relatively low electron temperatures at low gas pressures.

The 1-D extended nanospace channels on a Si wafer were fabricated by RIE. After the access holes for solution introduction were opened by KOH wet etching, the fabricated Si wafer was sealed with a cover glass substrate by an anodic bonding method at around 350 °C for 30 minutes, with an applied voltage of 800V.[7] Through a couple of dual-photolithography steps, using a stepper and contact aligner, and inductively coupled plasma ICP-RIE apparatus, nanopillars could be realized as extremely high aspect ratio (> 50:1) arrays, with various geometries, inside microchannels on a Si wafer.[8]

However, this photolithography-based fabrication method is difficult to apply for the 2-D nanospace patterning that is substantially smaller than the diffraction limit. One of the most powerful techniques for making complex and well-defined 2-D nanopatterns without a mask is electron beam (EB) lithography. A primary factor in EB lithography is the exposure dose (C/cm^2). Examples of EB resists include ZEP520A, ZEP7000 (Zeon Corp.), and OEBR-1000 (Tokyo Ohka Kogyo Co., Ltd.) as positive resists and SAL-601 (Shipley Corp.) as a negative resist. The characteristics and kinds of typical resists are shown in Table 3.1. The advantages of the ZEP series are the high resolution and good dry etching resistance. In particular, ZEP520A is characterized by a far higher resolution than ZEP7000, while ZEP7000 allows for a much thicker spin coat than ZEP520A. Figure 3.3 shows

Table 3.1. Relationship between various nuclear magnetic resonance (NMR) relaxation rate measurements and obtained information for examining molecular dynamics.

	ZEP520A (Zeon Corp.)	ZEP7000 (Zeon Corp.)	OEBR-1000 (Tokyo Ohka)	SAL-601 (Shipley Corp.)
Materials	Methoxy benzene	Diethylene glycol	Methacrylate resin	Novolac resin
Resist types	Positive	Positive	Positive	Negative
Changes of resist	Binary	Binary	Analog	Binary
Exposure doses	20~50 μC/cm^2	1~5 μC/cm^2	50 μC/cm^2	120 μC/cm^2
Properties	• High resolution • Slightly low sensitivity	• Slightly low resolution • High sensitivity	• High resolution • Low etching resistance	• High resolution • Low sensitivity/ low stability

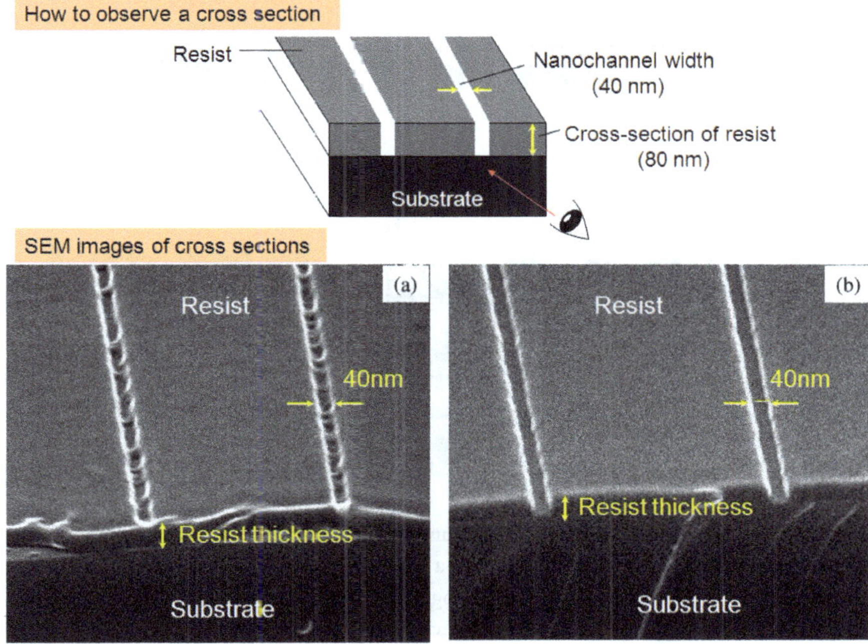

Figure 3.3. Schematic illustration of how to observe a cross section of a drawing. EB resist on a glass chip and SEM images of cross sections of EB resists in the case of (a) insufficient exposure and (b) sufficient exposure.

the patterning results of a fused-silica glass substrate coated with a ZEP520A resist after development. In the case of insufficient exposure, the fine patterns on the EB resist were impossible to draw and the EB resist itself remained on the surface, while, in the case of the well-suited EB exposure, the underlying glass surface could be bared. OEBR-1000 is an analog resist, in which the changes in the chemical structure depend linearly on the exposure dose, and it differs from a binary resist with exposure thresholds, such as the ZEP series. The OEBR-1000 has poor sensitivity and poor dry etch resistance, but can be well adhered to almost any surface and gives very reproducible results. On the other hand, the SAL-601 negative resist shows high resolution, like the ZEP520A, high contrast, and moderate dry etching selectivity. Since the SAL-601 resist's stability is quite poor, due to

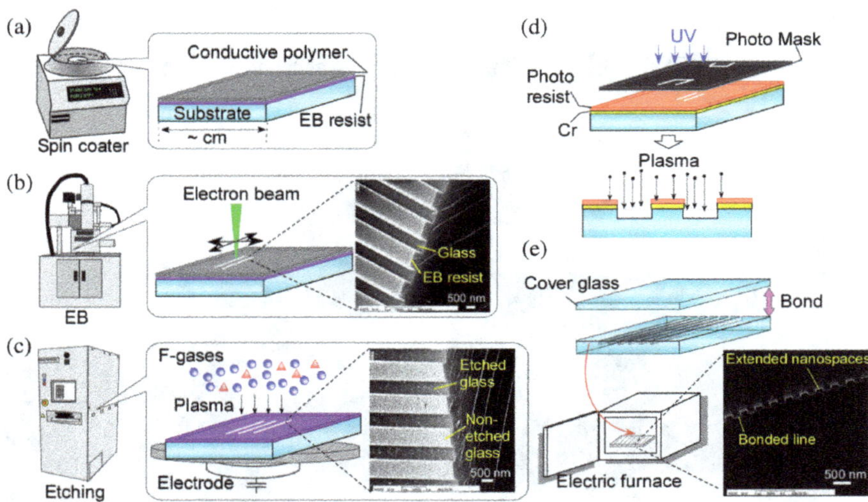

Figure 3.4. Typical fabrication procedures and their results: (a) Spin coating of EB resist and conductive polymer layers onto a substrate, (b) Drawing nanopatterns onto the layers on a substrate using EB lithography and scanning electron microscopy (SEM) image of drawn nanospace channels (540 nm width; 440 nm depth; 2 μm pitch) on a glass substrate, (c) Plasma etching of the nanopatterned surface and SEM image of the etched nanospace channels in a section, (d) Using photolithography and plasma etching for fabrication of microchannels, (e) Thermal bonding and SEM image of fabricated nanospace channels in section after thermal bonding.

the effects of temperature and weak adhesion to the surfaces, this is more suitable for patterning isolated features rather than dense ones.

Typical fabrication processes and their results are shown in Figures 3.4 and 3.5, respectively. After electron beam resists and conductive polymers were coated onto a substrate by a spin coater, the resists were pre-baked on a hotplate at various temperatures (typically ~180 °C). EB lithography allows the nanopatterns to be drawn directly onto the resist coated on the substrates. The drawn nanopatterns were developed in o-xylene and rinsed in 2-propanol, and then etched using an ICP-RIE apparatus with a mixture of gases consisting of CF_6/CHF_3 and CHF_3/O_2, or using a FAB apparatus with CHF_3 gas.[9–13] As a result, straight or Y-shaped 2-D extended nanospace structures could be obtained on the fused-silica substrate at etching rates

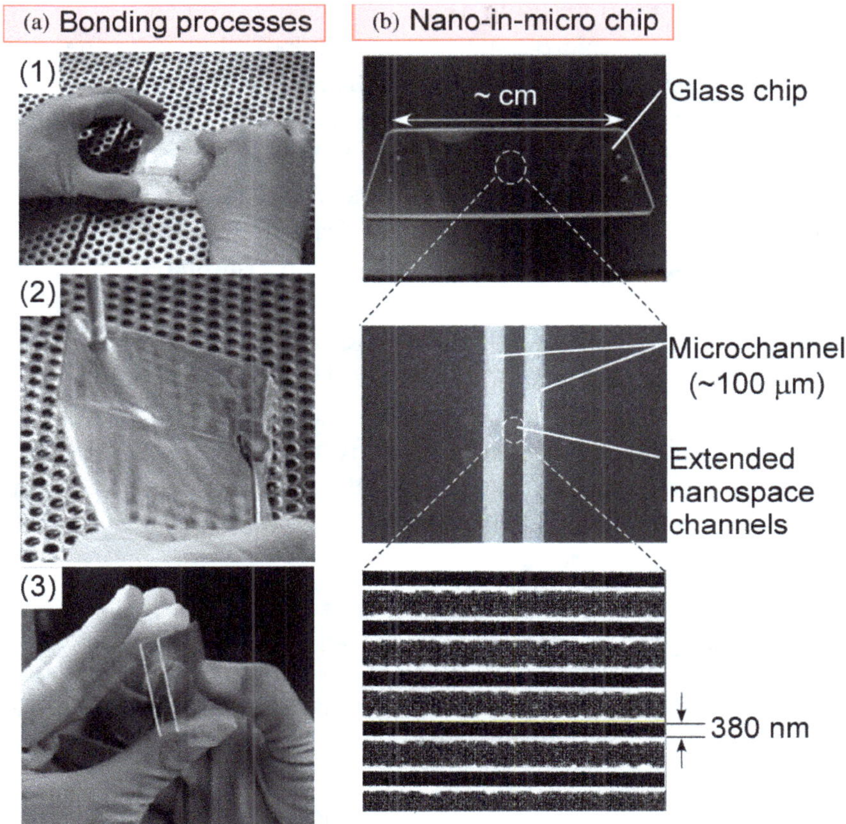

Figure 3.5. (a) Pictures of glass bonding processes: 1. In mechanical polishing, after the holder was fitted to the glass substrate, the substrate was polished thoroughly with melamine foam, 2. The substrate was washed repeatedly in flowing water after immersion in piranha solutions, 3. In aligning substrates by hand, the substrate was placed between the fingers, and two substrates were slowly brought close together (Adapted from Ref. 12), (b) Photographs and SEM images of the enclosed nano-in-micro structures on a chip.

of over 10 nm min^{-1}. Moreover, after a thin Cr layer and photoresist layer were deposited and coated on the fabricated substrate, respectively, the U-shaped microchannels for introducing solutions were fabricated by photolithographic patterning and plasma etching, or sandblast etching methods. The inlet holes were then pierced through the fabricated substrate using a diamond-coated drill. After the substrate with nano-in-micro structures was rubbed with melamine resin, including a surface-active agent, the substrate was washed repeatedly in o-xylene, dimethyl sulfoxide (DMSO), ultra-pure water, and a piranha solution (H_2SO_4:H_2O_2 = 3:1). By bonding with a cover glass in a vacuum furnace at 1080 °C, or a polydimethylsiloxane (PDMS) cover glass at room temperature, the fabrication of the enclosed nano-in-micro channels on a chip could be completed. Detailed information concerning the various bonding processes is available in Section 3.3. Nanopillars 100 nm wide and 1000 nm deep were embedded in a quartz-made microchannel using the fabrication techniques of EB lithography and NLD, with a mixture of CF_4 and C_3F_8 gases.[14] The high aspect ratio resulted from the utilization of Cr and Ni films as etching masks. The fabricated nanopillars-in-microchannel quartz substrate was thermally laminated with bare quartz at 1100 °C for 3 hours without pressurizing.

3.1.2.3. *Other lithographic techniques*

In order to realize the fabrication of minimal 2-D extended nanospace structures in the sub-100 nm range on a substrate, the light source needs to be shifted to wavelengths less than UV, such as deep UV with a wavelength in the 100~200 nm range, including excimer lasers (e.g. KrF; 248 nm, ArF; 193 nm) and Nd:YAG lasers, operating at the high-harmonic wavelength (3rd harmonic: 355 nm, 5th harmonic: 213 nm). The fabrication of a self-enclosed 2-D nanochannel was reported using the excimer laser melting method.[15] After a Si mold with 2-D nanochannels was fabricated by nanoimprint lithography (NIL) and RIE, the nanospace channels were exposed to an excimer laser. This melting phenomenon led to the reshaping of the surface on top of the nanochannels and joined the neighboring surfaces together. The channel sizes differed depending on the laser fluencies and melting times.

Interferomic lithography is a maskless technique based on the interference of two or more coherent beams. It makes it possible to fabricate nanopatterns inexpensively and quickly over large surface areas. The exposure source is the frequency-tripled (355 nm) output of a Nd:YAG laser. The laser beam is expanded, illuminating a right-angle assembly, containing a mirror and a vacuum chuck to hold the Si sample. The samples are etched using RIE with a mixture of O_2 and CHF_3. As a result, parallel 2-D extended nanospace channels, with widths of 100 nm and depths of up to 500 nm, are produced on a Si wafer.[16]

The utilization of grayscale photolithography draws a lot of attention to the fabrication of novel nanofluidic devices embedded with 3-D nanostructures. In this process, UV light, with different grayscale shade patterns, is transmitted through the photomask onto a photoresist coated on a surface. The amount of exposure can be determined by the amount of light permitted through the shade photomask, unlike conventional photolithography using monochromatic photomasks. After the grayscale photolithography the 3-D nanochannels, with a variety of depths in the range 10–100 nm, can be transferred onto a fused-silica substrate using RIE with a mixture of gases, O_2/CHF_3.[17,18]

3.1.2.4. *Direct nanofabrication*

The focused ion beam (FIB) method, in which nanostructures can be milled onto various substrates by the collision of ion beams such as Ga^+, has realized one-step, direct fabrication. In particular, minimal nanopatterns can be etched within very narrow areas on a substrate without any lateral scattering of ion beams. Using FIB, 2-D extended nanospace channels, with 100 nm scale dimensions (width and depth), have been milled onto <100> silicon, or fused-silica, substrates. The milled sizes were controlled through the variation of beam current. The substrates with 2-D extended nanospace channels were then bonded with a PDMS cover substrate or a fused-silica substrate, consisting of U-shaped microchannels for introducing solutions, respectively. These were selected as nano-in-micro structures.[19,20] The FIB milling technique also enabled 10 nm scale, 2-D nanochannel structures to fabricate on a silicon nitride (Si_3N_4) membrane, used as dielectric material. These nano-in-micro structures, containing 2-D nanochannels and microchannels, are utilized as manipulation devices for single DNA molecules.[21] Recently, it has been shown that 3-D nanofluidic channels, with 100 nm scale diameters, can be produced directly inside a glass substrate by using a femtosecond laser, since the laser focus can move freely inside transparent glass substrates. The glass substrate, with 3-D nanochannels, and a PDMS substrate, with reservoirs or microchannels, can then be bonded to each other.[22,23]

3.1.3. *Surface machining techniques*

The advantages of surface machining processes are that the extended nanospace channel structures can be fabricated without expensive, top-down nanolithography. As shown in Figure 3.1, after the bottom structural layer and sacrificial layer are deposited and patterned on a substrate, the top structural layer is deposited on top of the sacrificial layer. By etching the sacrificial layer, which is sandwiched between the top and bottom structural layers, a nanochannel can be successfully enclosed within a substrate. The size and shape of the nanochannels depend strongly on the thickness of the sacrificial layers.

3.1.3.1. *Utilization of polysilicon as a sacrificial material*

Silica compounds such as polysilicon, silicon nitride (Si_3N_4), and SiO_2 are generally utilized as sacrificial materials. [24–26] They are alternately sputtered onto a substrate by the low-pressure chemical vapor deposition method (LP-CVD), and etched through wet or dry etching processes. After the $Si_3N_4/SiO_2/Si_3N_4$ sandwich structures are formed, the silicon water, with sacrificial layers, is immersed in a HF solution. Since SiO_2 has high etching selectivity against Si_3N_4, only the SiO_2 layer is partially wet etched, and the nanochannels can be formed and enclosed within a Si_3N_4 layer. The layer materials are deposited and patterned onto a substrate in the order of: the bottom structural layer (Si_3N_4 or SiO_2), the sacrificial polysilicon layer, and the top structural layer, on a substrate using the LP-CVD method. The polysilicon sacrificial layers can then be wet etched selectively with immersion into an etchant, such as TMAH and KOH. This method is suitable for fabricating 1-D nanochannels with nanometer-depth, due to its high etching rates (approximately 1 μm min^{-1}) of polysilicon in 20–30% KOH solution, and a few per cent TMAH solution, at about 80 °C.

Schoch and coworkers showed that an amorphous silicon (aSi) material is helpful as not only a nano-sized spacer between the top and bottom substrates, but also an assistant medium for bonding processes.[27] By only sputtering a Si sacrificial layer, 1-D nanochannels with 50 nm height could be fabricated on a Pyrex glass substrate possessing microchannels.

Several reports demonstrated that a sacrificial layer could be removed using dry etching methods. Polysilicone and SiO_2 were utilized as the sacrificial layer and the structural top and bottom layers, respectively, using the LP-CVD method, and isotropically etching the polysilicone with Xenon difluoride (XeF_2) gas for 1.5 hours.[28] SF_6- or CHF_3-based plasmas were also exposed to an aSi sacrificial layer sandwiched between top and bottom structural layers of SiO_2 by a high-density ICP apparatus. The etching of the aSi layer proceeded depending on the exposure times (etching rate: ~100 nm min^{-1}).[29,30]

On the other hand, a novel fabrication concept has been recently developed that combines lithography, etching, and thermal oxide growth of sacrificial materials.[31] This fabrication process is shown in Figure 3.6(a). After the extended nanospace channels, with about 500–1000 nm dimensions, were patterned by standard photolithography, the patterned surface was etched by an anisotropic wet etching process due to KOH or a deep-RIE process for achieving deep-trenched etching on the Si substrate. Since the deep etching processes led to greater channel width, the thermal oxidation process was applied to grow an oxide sacrificial layer for narrowing the channel width below 10–100 nm. A structural layer was then deposited, by non-conformal plasma-enhanced chemical vapor deposition (PECVD), to seal the nanospace channels. As a result, enclosed 2-D extended nanospace channels, with a very high aspect ratio (50 nm wide at a depth of 40 μm), can be built onto the Si substrate (see Figure 3.6(b)). This fabrication method is compatible with complementary metal oxide semiconductor (CMOS) processes.

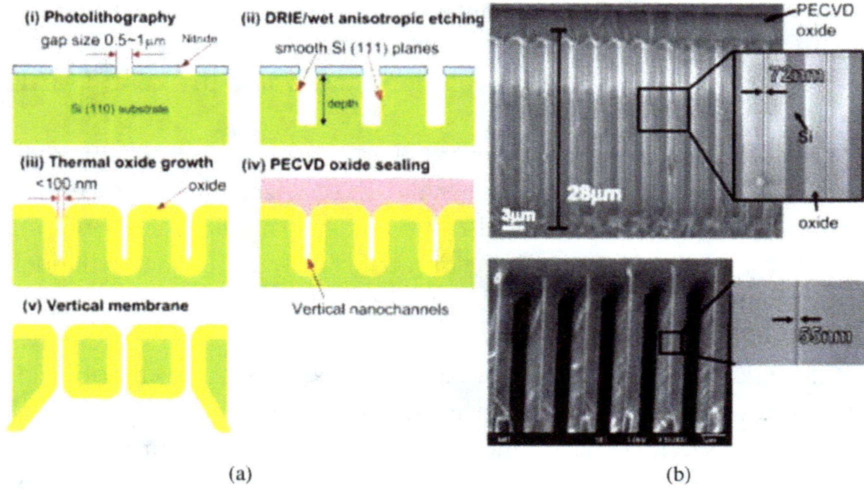

Figure 3.6. (a) Schematic illustration of vertical extended nanospace channel fabrication processes: (i) photolithography, (ii) deep-RIE or wet anisotropic etching, (iii) thermal oxide growth of sacrificial layers, (iv) PECVD, (v) backside etching of the Si substrate, (b) Cross-sectional SEM images of vertical extended nanospace channels. The channels are etched by KOH etching, have a depth of 28 μm, and are completely sealed by depositing 3 μm thick PECVD oxide (Adapted from Ref. 31).

3.1.3.2. *Utilization of metals and polymers as sacrificial materials*

Similar methods, which combine the deposition of sacrificial layers and the selective etching of these layers, have been implemented by using other materials. Firstly, a 100 nm thick layer of aluminium was deposited on a silicon wafer coated with photosensitive polyimide, using sputtering methods.[32,33] After the silicon wafer was patterned by photolithography and standard aluminum etching processes, the top of the patterned, sacrificial aluminium and bottom polyimide layers were deposited due to the polyimide layer. Since only the sacrificial aluminum layer between the top and bottom polyimide layers can be removed using aluminium etchant, the formation of 1-D extended nanospace channels was easily realized. The repeated deposition of aluminium, SiO_2, and photoresists as sacrificial layers, and its patterning, can establish unique, segment-type enclosed 1-D nanochannels with differing heights on a substrate. Further, the crystalline silicon germanium (SiGe) was deposited as a sacrificial layer on the substrate, followed by chemical vapor deposition (CVD) and patterning.[34] The SiGe sacrificial layer could then be selectively wet-etched to form 10 nm-sized extended nanospaces. By depositing the SiO_2 layer covers, the nanospaces are easily sealed and utilized as nano array devices using capillary filling control.

Combining different metals made it possible to establish a novel electrochemical device, consisting of nanoelectrodes and nanochannels embedded in a substrate.[35,36] SiO_2, chromium (Cr), and platinum (Pt) were used as structural, sacrificial, and electrode layer materials, respectively. The sacrificial Cr layer was selectively etched by immersing the device in chromium etchant, and nanochannels surrounded with a Pt layer were created.

A fabrication technique using thermal-degradable polymers as a sacrificial material has also been developed.[37,38] A polynorbornene (PNB) polymer structure was patterned on a silicon wafer, followed by the deposit of SiO_2, photoresist, and Cr layers and its patterning. A SiO_2 structural layer was then deposited on top of the patterned PNB and bottom silicon wafer, using the CVD method. Since the PNB embedded in the spaces between the top SiO_2 layer and bottom silicon wafer could be decomposed through increasing temperatures above 400 °C, because of the glass transition temperature of PNB (~350 °C), 100 nm-sized 2-D nanochannels formed together with the sealing.

3.1.4. *Imprinting and embossing nanofabrication techniques*

NIL refers to a technique in which a precursor resist is cast against a mold, and can achieve resolutions beyond the limitations of light diffraction or beam scattering. The nanopatterns on the mold can be easily duplicated into the resist in large quantities, due to repeated utilization of the mold. However, NIL has the fault that a mold itself has no choice but to fabricate using either bulk or surface nanomachining processes. The NIL methods include both thermal NIL, and photo-curing NIL. Their general fabrication steps are illustrated in Figure 3.1.

In the thermal NIL process, thermoplastic materials such as polymethylmethacrylate (PMMA) and polycarbonate (PC) polymer films are used as a cast for the silicon or glass templates containing nanostructures.[39] The deformation of these materials can be generated according to changes in temperature above and below the glass transition temperature (T_g) of the materials, so that both the Young's modulus and the viscosity of the materials are swung by several orders of magnitude with changes in temperature. A mold with extended nanospace patterns, fabricated on a SiO_2 layer deposited on a Si, or glass, substrate as shown in Figure 3.7(a), was impressed to deform the PMMA resist coated on a substrate at a temperature of 175 °C

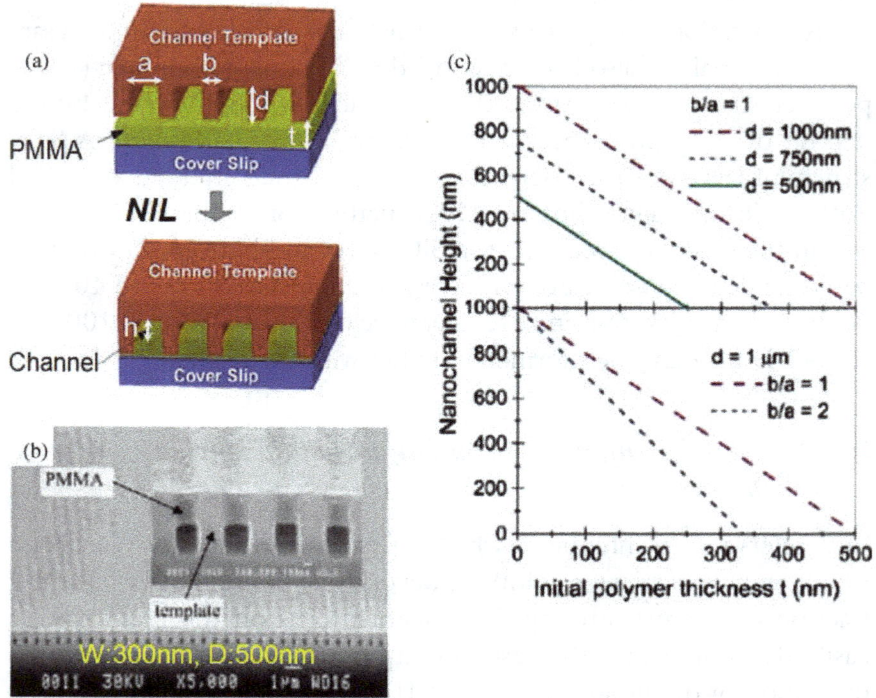

Figure 3.7. (a) Fabrication procedures for thermal NIL of PMMA polymer, (b) SEM images of extended nanospaces enclosed between PMMA polymer and template, (c) Resist condition and mold size dependences for extended nanospace heights (Adapted from Ref. 39).

and a pressure of 5 MPa for 5 minutes. These conditions were used due to the T_g values of PMMA, which range from 85–165 °C. When the mold and PMMA substrate were cooled down to ambient temperature, enclosed 2-D extended nanospace channel arrays, with 100 nm-sized width and depth, could be achieved (see Figure 3.7(b)). The height of the extended nanospaces could be controlled by the polymer thickness and/or the pitches and sizes of the molding with nanochannels (see Figure 3.7(c)).

A piece of uniform silica nanowire, that was fabricated using a flame-heated fiber drawing device on a PC substrate, was chosen as a template for embossing the nanospaces.[40] Since the silica nanowire was sandwiched between a PC substrate and a glass substrate at above 100 °C and a few MPa, the silica nanowire was embedded in the PC substrate. The 2-D nanochannel emerged on the PC substrate through the removal of the silica nanowire under a HF solution. The PC substrate was then sealed with a PDMS microchannel substrate for the realization of nano-in-micro structures. After a SU-8 master with 2-D nanopatterns (width: 580 nm, depth: 130 nm) was fabricated by EB lithography to within an accuracy of a few per cent, PDMS prepolymer was poured over the SU-8 master and cured at 70–120 °C for the fixed time. By aligning the fabricated PDMS substrate and a boronosilicate glass substrate containing microchannels under an optical microscope, the 2-D nano-in-micro structures could be reversibly sealed together.[41]

Thermal NIL, that requires high temperatures and pressures, often occurs due to the mismatch between a mold and an imprinting substrate, because of its thermal expansion. Utilization of liquid precursors that can be cured by UV irradiation at ambient temperatures is useful as an alternative method, called "photo-curing NIL". An imprinting mold, which has a single, sub-20 nm width and depth, and cm length, was pressed into the top layer of UV-curable resist (NXR series, Nanonex Corp.) coated on a SiO_2 substrate using a nanoimprinter.[42] The nanopatterns on the mold could then be transferred to the resist on a substrate. Following that, the resist was cured by UV irradiation and the cured resist was removed from the mold. The imprinted SiO_2 substrate was then etched by CF_4-based RIE. As a result, a minimal 2-D nanochannel, with 10 nm scale width and depth was successfully transferred onto a SiO_2 susbstrate. The quartz mold with nanostructures was pressed into a thick photo-curable photoresist (SU-8) coated on a quartz substrate, and then UV-irradiated to cure the imprinting SU-8 nanostructures.[43] When the imprinted chip and a SU-8 spin-coating quartz substrate were pressed together under UV exposure, an enclosed 2-D nanochannel with 250 nm width and depth was achieved due to cross-linking SU-8 polymers. A polyethylene glycol (PEG) polymer, such as PEG diacrylate (PEG-DA), was also used as a UV-curable resist for nanofabrication. The PEG-DA resist was drop-dispensed onto a Si mold, fabricated using photolithography or EB lithography depending on the designed sizes, and a supporting polyethylene terephtalate (PET) film was pasted on top of the PEG-DA surface, coated on a Si mold. After the sample was exposed to UV light for a few minutes, the PEG-DA material was peeled off the Si mold. When the PEG-DA material came into contact with the gold or silicone substrate, the enclosed 2-D nanochannels could be established.

3.1.5. *New strategies of nanofabrication*

3.1.5.1. *Non-lithographic techniques*

Novel approaches taking advantage of non-lithographic processes, based on physicochemical deformation of polymer substrates, have also been demonstrated.[44–46] Concretely, linear nanoscale cracking structures are formed by stretching an oxidized PDMS substrate using controllable stretching apparatuses. After the nanocracking PDMS substrate is imprinted onto a UV-curable epoxy resin, the imprinted resin with triangle-shaped 2-D extended nanospaces is repeatedly used as a mold, against a flat PDMS substrate. The imprinted PDMS substrate makes it possible to seal it with other PDMS microchannel substrates through pressurizing at room temperature. If the nanocracking PDMS substrate is directly exposed to oxygen plasma and then bonded to a flat PDMS substrate, PDMS nano-in-micro structures containing 2-D extended nanospace channels and microchannels are constructed.[47,48] The film is then placed into a closed chamber saturated with toluene vapor at room temperature. Upon exposure to the vapor, the internal osmotic pressure induced by the preferential swelling of the unmodified polystyrene results in a net compressive stress in the system. When the stress exceeds a critical threshold value, surface wrinkles begin to form with the pattern morphology exhibiting a clear dependence on the length of UVO exposure. A rich variety of patterns can be observed by simply varying the exposure time between 2 and 40 minutes; these include flowerlike patterns, spokes, targets, labyrinths, and dots.

Microchannels are embossed from a patterned silicon wafer mold and thermally bonded to a covering PC substrate.[49] When the embossed PC chip is stretched right and left from the center of the heating region, at above critical T_g, some of the microchannels in the PC chip are lengthened. This thermal stretching enables the microchannels to reduce in size, in a controlled manner, until they form 400 nm-sized nanospaces, with circular or elliptical cross sections.

3.1.5.2. *Hybrid-material techniques*

Some researchers have proposed a simple fabrication method for a hybrid-nanofluidic chip that is composed of a microchannel substrate and nanoporous membrane. Demonstration of the two fabrication approaches is accelerating; one is a nanoporous-junction method and the other is a nanocapillary array method. In the former case, the Nafion resin, with 5 nm pore size, can act as a separator at a nanoscale level, because the Nafion solution is introduced by mechanically cutting junctions across the microchannels on a PDMS substrate through capillary force, with the Nafion resin then embedded in the junction.[50,51] On the other hand, in the latter case the 10–100 nm scale nanocapillary array membrane (NCAM), such as polycarbonate, is aligned and sandwiched between two microchannel substrates and bonded together using adhesive.[52,53] Since the flow of solutions results from capillary action into the NCAM from the upper microchannel to the lower microchannel, the application possibilities are limited, so far.

3.1.6. *Combination of lift-off and lithography*

Recently, a nanogap electrode has been embedded in a portion of a top-down fabricated 2-D extended nanospace channel.[54] After the 2-D nanochannel, with 10 nm width and depth, was fabricated on a SiO_2 substrate by NIL and RIE etching processes, the NIL process was performed again on the substrate with the 2-D nanochannel. In this case, the imprinting was carried out in the direction perpendicular to the fabricated 2-D nanochannel. After that, Ti/Au metals were sputtered on top of the imprinted resist on a substrate via the shadow evaporation method, with two symmetric tilted angles. By using a lift-off, the Ti/Au metallic nanowire could form in a 2-D nanochannel on a substrate. As a result, the nanowire-in-nanochannel structure was successfully realized by sealing with a cover plate.

Nano-in-micro structures on a chip also made it possible to realize cell culture and its patterning.[55] Briefly, gold stripe nanopatterns with extended nanometer-sized dimensions were lifted-off on a quartz glass substrate using EB lithography and metal sputtering methods as shown in Figure 3.8(a). The fabricated substrate was then bonded with a PDMS, or bare glass, substrate embedded with a microchannel. By introducing solutions containing self-assembled monolayers, such as octadecanethiol and/or alkanethiol, into the nano-in-micro structures, only the monolayers were patterned onto the gold nanopatterns. Since mouse fibroblast NIH/3T3 cells were attached onto only the nanopatterned surface, they could be aligned better along the direction of nanopatterned lines and changed depending on the nanopatterned sizes (see Figure 3.8(b)). Such cell direction control was observed on the nanopatterned surface, invoking metal stripes and monolayers, but was not observed in the case of glass-etched nanostructures.

3.2. Local Surface Modification

Extended nanospace shows more orders of higher surface-to-volume ratio than does microspace, and the surface properties play significant

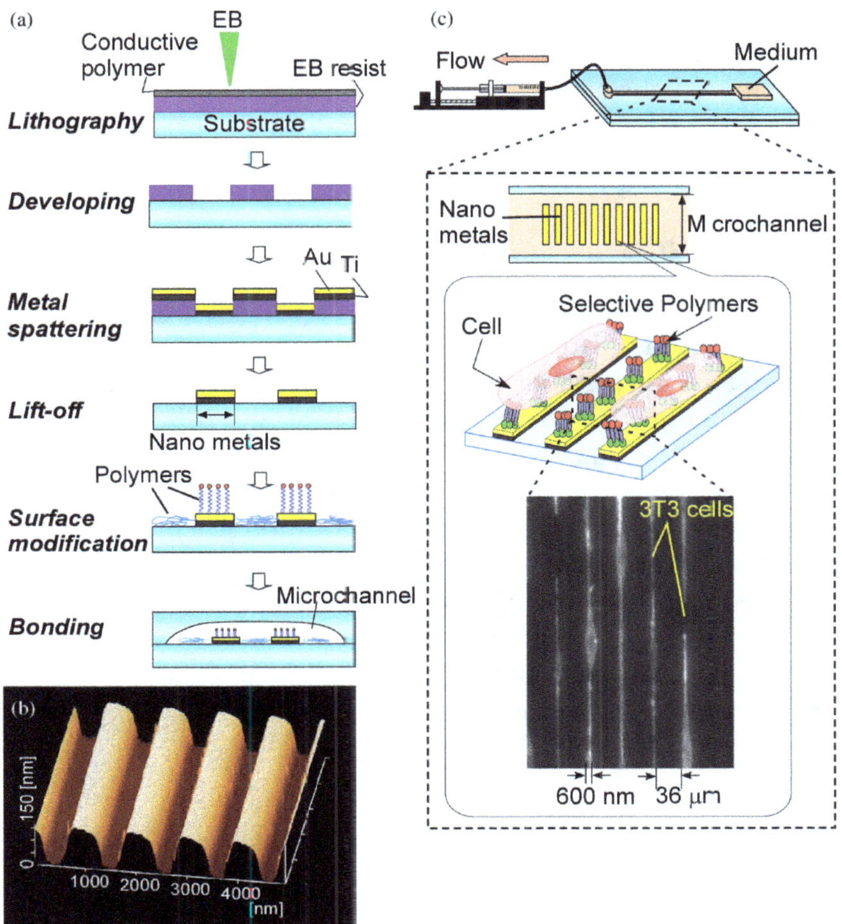

Figure 3.8. (a) Fabrication scheme of metal nano-in-micro structures on a glass chip, (b) Atomic force microscope (AFM) images of the patterned metal nanostructures, (c) A micrograph of 3T3 cells on the patterning metal extended nanospace structures (600 nm width, 50 nm height, and 36 μm pitches) in a microchannel (Redrawn from Ref. 55).

roles for fluidic and reaction controls. For example, local hydrophobic and hydrophilic patterning can control the two-phase multiphase flow, and the local immobilization of the proteins and DNA can be utilized for single bio-molecule capture and analysis. In some cases, the surface properties significantly affect the liquid properties. Therefore, control of surface properties in the local area is essential for integrated chemical processes in extended nanospace. Here, we show several reports on micro and extended nanosurface patterning methods.

3.2.1. *Modification using VUV*

Vacuum ultraviolet (VUV) light lithography has previously been applied for surface patterning. VUV lithography is an extremely efficient and useful method for fabricating a patterned organosilane layer. It can directly decompose most organic materials with its high proton energy, eliminating the need for a photoresist and photosensitive materials. Here, we introduce a method for protein micropatterning. Figure 3.9 shows a fabrication scheme of the VUV irradiation. VUV etching is carried out using a vacuum chamber to fabricate a patterned layer.[56] The substrates are placed in a chamber, which is almost a vacuum, and irradiated with VUV light through both a thin metal mask and a quartz glass plate fixing the mask to the substrate. In some cases, some weight was put on the glass to ensure contact between the mask and the substrate. This way, only

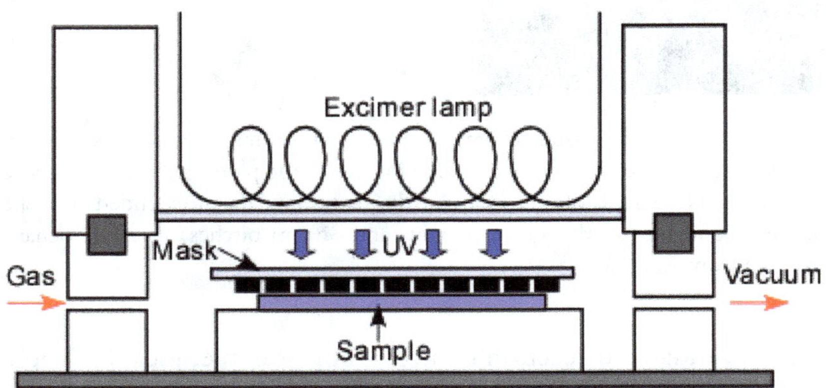

Figure 3.9. Schematic illustration of a VUV irradiation system (Redrawn from Ref. 56).

proteins irradiated with VUV are decomposed, and the micropatterning is completed.

This VUV-based patterning is used for several applications, such as cell patterning.[57] Cell patterning, a method for spatially controlling cell adhesion and growth on substrates, is important for a wide range of applications, such as basic biological study, cell-based biosensors, and tissue engineering.[58,59] In order to realize cell patterning, non-cell adhesive materials must be patterned on the cell culture surfaces. Here, PEG was used which has been shown to significantly inhibit nonspecific interactions with various proteins, platelets, and cells; hence, the polymers forming possess excellent biocompatibility when using both traditional and biomedical applications. By using the VUV patterning described above, the PEG is patterned and the cell is also patterned on the desired area.

3.2.2. *Modification using an electron beam*

The electron beam (EB) is a powerful tool for surface modification in the nanometer scale. Here, we introduce a simple process for nanopatterned cell culture substrates, by direct graft-polymerization using an EB lithography system requiring no photomasks or EB-sensitive resists.[60] The compound N-isopropylacrylamide (IPAAm) was locally polymerized and grafted directly by EB lithographic exposure onto hydrophilic polyacrylamide (PAAm) grafted glass surfaces. The size of the surface-grafted polymers was controlled by varying the area of the EB dose, and a minimal stripe pattern with a 200 nm line; width could be fabricated on the surface. On the stripe-patterned surfaces, above the lower critical solution temperature (LCST), the cells initially adhered and spread with an orientation along the pattern direction. The magnitude of the spreading angle and elongation of adhered cells depended on the pattern intervals of the grafted PIPAAm. When the culture temperature was lower than the LCST, cultured cells detached from the surfaces with strong shrinkage along the pattern direction, and sometimes folded and became parallel to the stripe pattern. This patterned cell recovery technique may be useful for the construction of muscle cell sheets with efficient shrinkage/relaxation in a specific direction and spheroidal 3-D cell structures, with application in tissue engineering and microfluidic cellular devices.

3.2.3. *Modification using photochemical reaction*

Photochemical reaction is an effective method for patterning chemicals in a closed space without destroying the other chemicals, like VUV. In this section, we introduce a cell patterning method in a closed microchannel, using UV light.[61] Figure 3.10 shows a schematic of the cell patterning. This technique focuses on the use of a modified 2-methacryloyloxyethyl phosphorylcholine (MPC) polymer, which is known to be a non-biofouling compound that is a photocleavable linker (PL), to localize cells via connection to an amine-terminated silanized surface. Using UV light illumination, the

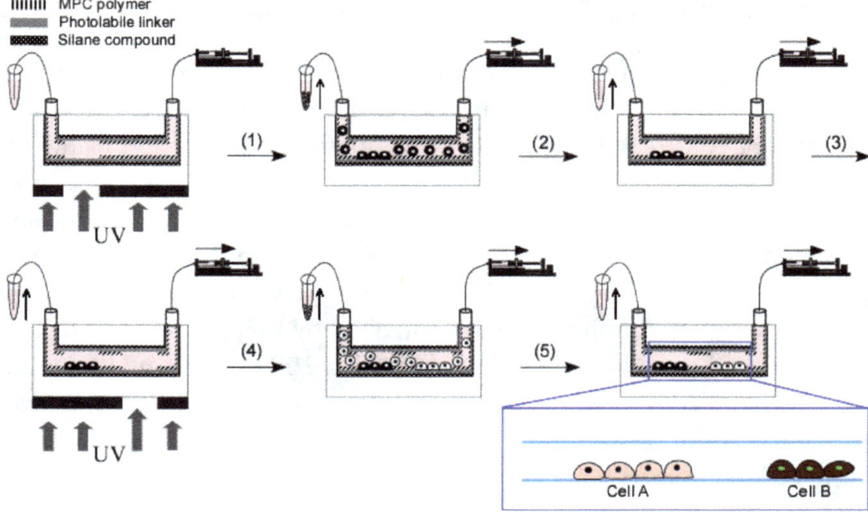

Figure 3.10. Schematic demonstration of two different types of cell patterning in a microchannel: (1) Introducing Cell A suspension, (2) Fresh medium was introduced using a microsyringe pump, (3) The microchannel was exposed to UV through a photomask, (4) Cell B suspension was introduced, (5) The microchannel was connected to a microsyringe pump. Cell A was immobilized in 2 cm and Cell B was immobilized in 7 mm along the microchannel. The distance between two patterns was 1 cm long (Redrawn from Ref. 61).

MPC polymer was selectively eliminated by photochemical reactions that controlled the cell attachment inside the microchannel. After selective removal of the MPC polymer through the photomask, MC-3T3 E1 cells, and vascular endothelial cells (ECs) were localized only on the UV-exposed area. In addition, the stability of the patterned ECs was also confirmed by culturing for two weeks in a microchannel under flow conditions. Furthermore, two different types of cells were cultured inside the same microchannel through multiple removals of the MPC polymer. ECs and Piccells, which are cell-based NO indicators, are localized in both the upper and lower streams of the microchannel, respectively. When the ECs were stimulated by adenosine triphosphate (ATP), NO was secreted from the ECs and could be detected by fluorescence resonance energy transfer (FRET) in the Piccells. This technique can be a powerful tool for analyzing cell interaction.

For fluidic control in extended nanospace, high pressure (0.1–1.0 MPa) is required for pressure-driven flow. In this case, strong bonding for substrates will be essential, but the conventional PDMS/glass bonding is weak in order to meet the high liquid pressure, and thus cannot be used in extended nanospace. The modification method using open channels, such as EB, AFM, scanning tunneling microscopy (STM), contact printing, and ink-jets, is therefore not appropriate because the bonding process utilizes high temperature or specific reagents, which remove the immobilized molecules. Due to this, local surface modification with light is one candidate for closed extended nanospaces. So far, there are only a few reports on the local surface modification in the extended nanospace utilizing light. Here, we show an example of local DNA immobilization in the extended nanospace using a photo-linker, as shown in Figure 3.11.[62] A photo-linker of PEG-Benzophenone was synthesized for this. Benzophenone reacts with the CH group in target DNA by photochemical reaction, and the target DNA is immobilized. PEG works as a protector against nonspecific adsorption. The hydrophilic property of PEG is also important for the introduction of liquid with moderate pressure because the Laplace pressure becomes so high (~MPa) in extended nanospace. The local DNA patterning was confirmed by fluorescence microscope after hybridizing

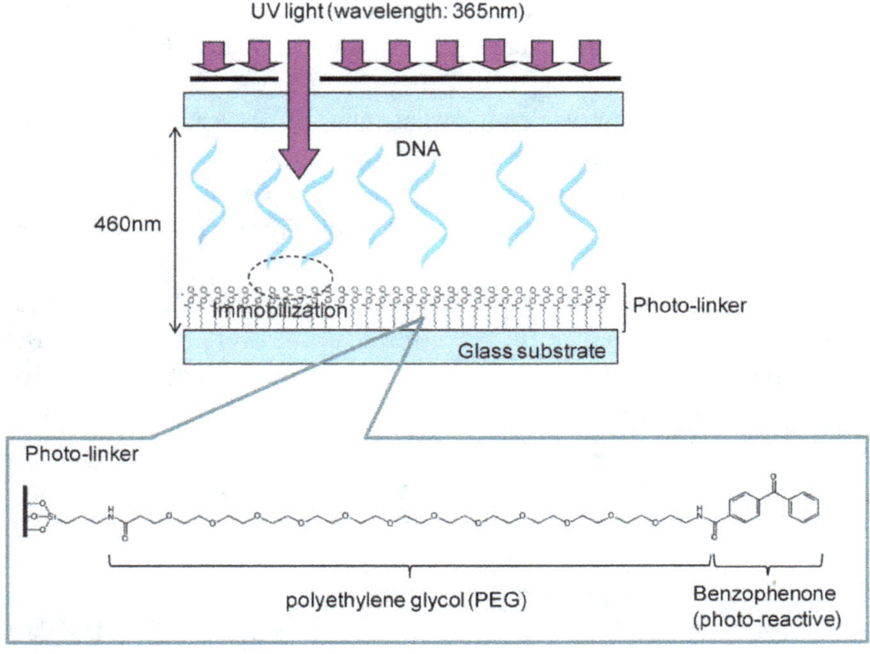

Figure 3.11. Local immobilization of DNA in extended-nanospace (Redrawn from Ref. 62).

with fluorophore-labeled, complementary DNA and also by investigating its cross reaction with non-complementary DNA.

More local immobilization methods are required for fluidic control, reaction control, and heat transfer control. As a primary objective in these methods, the spatial resolution should be improved. The spatial resolution is limited to the optical diffraction limit, which depends on the wavelength. The size of the extended nanospace is comparable to, or smaller than, the optical diffraction limit. This new concept will be essential in overcoming these problems. Other than these technologies, there are many molecular patterning methods such as inkjet printing,[63] lithography,[64,65] microcontact printing,[66] or nanoimprinting.[67,68] These can realize exact patterning, but they are difficult to apply in closed spaces. Development of the mild bonding method is another key technology using these general modification methods.

3.3. Bonding

3.3.1. *Introduction*

Wafer-to-wafer bonding processes provide a means for additive microfabrication, which complements the subtractive bulk micromachining processes. When applied to the fabrication of micro- and nanofluidic systems, these processes extend the scale and material properties of the structures obtainable through thin film deposition. This section provides an overview of the common wafer-scale bonding processes, which enable the fabrication of micro- and nanofluidic systems, including direct bonding, plasma activated bonding, and anodic bonding. These methods are the subject of many books and review articles, and the reader is directed to these publications for further details.[69–75] The purpose of this section is not to provide a thorough review of these methods, but rather to serve as a general introduction to each method and describe the basic theory needed to understand the underlying mechanisms.

3.3.2. *Wafer bond characterization methods*

The bonding methods described in this section share a common set of tools used for bond characterization. Bonded wafers are characterized in terms of both bond strength and uniformity, where strength can refer to either the force or energy per unit area required to separate the bonded surfaces. Uniformity can be non-destructively monitored as part of the bonding process, using imaging methods such as optical transmission, scanning acoustic microscopy, and X-ray topography. Optical transmission is easily carried out by illuminating the bonded pair with a white light or infrared source, and imaging the bond interface with a charge-coupled device (CCD) camera equipped with the appropriate filter. Using this method, separations as small as one quarter of the source wavelength will appear as interference fringes called "Newton rings", which indicate a void in the bond interface. However, optical transmission can only be applied when there is a range of wavelengths corresponding to photon energies that lie within the bandgap of the materials to be bonded, and for which the CCD has sufficient sensitivity. On the other hand, scanning acoustic microscopy identifies voids based on their elastic properties, and hence can be applied to any material combinations.[76] In this method, a piezoelectric transducer creates a travelling plane wave inside a fluid-filled cavity, which is focused onto one surface of the bonded wafer pair using an acoustic lens that is typically made from a material such as sapphire, in which the speed of sound is almost ten times larger than in water. The reflected wave's magnitude, which depends on the local elastic properties of the bonded pair, is transduced into an electric signal by the same acoustic lens and transducer. X-ray topography provides high-resolution images of lattice defects in the crystal structure of the bond interface, but can only be applied to single-crystalline materials.[77] Several destructive characterization techniques are also available for measuring bond uniformity, including scanning electron microscopy (SEM) and transmission electron microscopy (TEM).[78] SEM is advantageous from the point of view of sample preparation, requiring only cleavage of the bonded pair. Compared to this TEM requires the mechanical grinding, polishing, and ion milling of a cross-sectional sample to a final thickness that allows transmission of the electron beam, usually less than 300 nm. Bond strength can be measured using either the crack propagation or tensile/shear loading techniques. The crack propagation, or double cantilever beam test, involves inserting a razor blade in between the bonded wafers and measuring the resulting length of delamination by optical imaging.[79] If both wafers have the same thickness and Young's modulus, the surface energy can then be estimated by the relationship $\gamma = 3Et_w^3t_b^2/32L^4$, where γ is the surface energy, E and t_w are the wafer's Young's modulus and thickness, respectively, t_b is the thickness of the blade, and L is the crack length.[80] While this method is very simple to carry out, the measured surface energy will depend on local variations in surface roughness and stored elastic strain, as well as the ambient relative humidity; hence averaging over multiple measurements is often required. Another measure of bond strength is the tensile or shear load needed to separate a bonded pair from having a well-defined contact surface area.[81] This is usually performed by die-sawing wafers into small pieces and gluing them to metal jigs, which interface with a commercial pulling tool. This method is advantageous in that both tensile and shear modes of failure can be tested, and it can provide more information for samples with very high bond strengths approaching that of the bulk materials, for which the crack propagation method can be impractical.

3.3.3. *Wafer direct bonding*

Wafer direct bonding, or fusion bonding, refers to the joining of two smooth, flat, and clean surfaces without the use of externally applied forces or intermediate adhesion layers. The development of wafer direct bonding methods has been driven primarily by the semiconductor industry's need for a versatile means of producing engineered substrates with combinations of crystalline materials that cannot be obtained through epitaxial methods. For example, silicon-on-insulator (SOI) substrates consisting of a thin single-crystalline silicon "device" layer (from several tens of nm to several hundreds of μm thick) separated from a bulk mechanical silicon layer by a buried silicon dioxide layer (from several hundreds of nm to several μm thick) are enabling the development of new devices for low-power electronics, optoelectronics, and silicon nanowire applications. SOI substrates are often produced using the so-called smart-cut process, whereby one single-crystalline silicon wafer is directly bonded to another, which has been thermally oxidized and implanted with hydrogen/helium ions to a specified depth.[82] The subsequent annealing of the bonded pair results in agglomeration of the implanted ions at the edge of the implanted region, crack propagation, and the transfer of the thin film. Other applications of direct bonding include the manufacture of highly multilayered 3-D integrated circuits,[83] as well as (bio)chemical sensors and actuators based on micro-/nano-electromechanical systems (M/NEMS),[76,84] the latter area being the focus of this section.

3.3.4. *Wafer direct bonding mechanism*

For chemical and biological M/NEMS applications, the most common processes require the bonding of glass materials, and hence this sub-section will deal mainly with the direct bonding of silicon dioxide surfaces. However, many of the basic principles discussed here apply to any hydrophilic materials, as we will see shortly. Bonding is initiated by bringing the two cleaned surfaces into contact at room temperature in atmospheric air, without any applied pressure. Intimate contact is usually initiated by gently pressing the wafers together at their centers. If the surfaces are sufficiently smooth, flat, and free of particle and chemical contaminants, the contact area will spontaneously spread out to cover the entire interfacial surface through a combination of attractive forces including van der Waals forces, and hydrogen bonds between silanol groups on the opposing oxide surfaces, as shown in Figure 3.12.[71] This spontaneous bonding is enhanced by the presence of thin films of water, in the order of a few monolayers thick, which extend the range of interfacial attraction.[85] Maszara *et al.* observed that, during spontaneous bonding of oxidized silicon wafers, intimate contact is more readily established if the wafers are exposed to ambient cleanroom air, for a duration of minutes to hours, or spraying them with deionized water, rather than

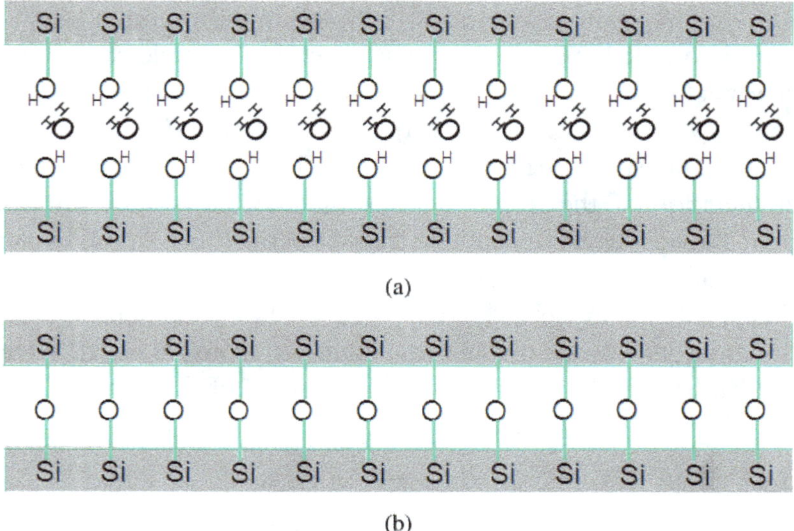

Figure 3.12. Schematic illustration of the principle of wafer direct bonding.

immediately contacting the wafers after removal from the furnace.[79] They also found that spontaneous bonding is either reduced, or completely eliminated, by exposing the surfaces to a vacuum, which confirms the importance of interfacial physisorbed water films for establishing intimate contact. In separate experiments, Lasky showed that if oxidized silicon wafers are brought into contact in a dry oxygen ambient at 1000 °C, spontaneous bonding does not occur, which was attributed to the complete conversion of silanol to siloxane on each surface, thus eliminating sites for bonding.[86] The spontaneous bonding of hydrophilic glass surfaces at room temperature is reversible, and the measured bond energies are consistent with the attractive forces expected for hydrogen bonds between silanol groups and bridging water molecules. Once intimate contact has been established, water removal is necessary in order to produce permanent siloxane bridges across the bond interface, according to the reaction

$$\equiv\text{Si-OH} + \text{HO-Si}\equiv \rightarrow \equiv\text{Si-O-Si}\equiv + \text{H}_2\text{O}.$$

Because water removal by diffusion is extremely slow at room temperature, covalent bonding is commonly achieved by annealing the bonded pair at an elevated temperature. The dependence of bond energy on anneal temperature and duration has been studied experimentally by Maszara *et al.* for 100 silicon wafers, on which a 300 nm wet thermal oxide was grown.[79] The bond energy kinetics were found to exhibit three distinct regimes. In the first regime, from room temperature to 400 °C, bond energy is time dependent for anneals up to one hour in duration, gradually increasing with temperature. This regime is characterized by the removal of physisorbed water, which is both temperature and time dependent (a more detailed discussion of reaction kinetics in this regime is provided in Section 3.3.6). These observations are consistent with the X-ray photoelectron spectrum and high resolution electron energy loss spectrum measurements of native oxide films on silicon, which indicate that hydroxyl groups are converted to silicon dioxide and hydride in the 400–500 °C temperature range.[87] In the second regime, comprising temperatures between 400 °C and 1100 °C, bond energies are found to be independent of time but continue to gradually increase with temperature. This behavior is explained in terms of the increased availability of bonding sites due to the elastic deformation of the wafers, supported by evidence that the mechanical thinning of one wafer increased the bond energy. In the third regime, in which annealing is performed between 1100 °C and 1400 °C, bond energy again becomes time dependent and highly sensitive to temperature, achieving a bond energy comparable to that of bulk silicon at 1400 °C. This behavior is attributed to plastic deformation of the surfaces due to the viscous flow of oxides at these temperatures and the resulting increase in contact area. This is consistent with Taft's observation that the refractive index of thermal oxides coincides with that of fused quartz above 1100 °C.[88] While these results provide a qualitative picture of the mechanisms involved in direct bonding of silicon dioxide surfaces, quantitative predictions of the surface energy dependence on temperature and time remain elusive.[85,89,90] In practice, the choice of annealing temperature may be limited by thermal budgets imposed by the presence of heterojunctions, such as metal contacts, for which thermal expansion mismatch or interdiffusion may be undesirable. Low temperature, direct wafer bonding, where annealing is performed at temperatures between room temperature and 400 °C, can produce high bond strength, but extremely long annealing times may be needed to reach the equilibrium value of bond energy, unless the surfaces are treated appropriately, as discussed in Section 3.3.6. In other cases, wafers have been successfully bonded at an intermediate temperature of 450 °C by employing glass materials having a lower glass transition temperature than thermal oxide, such as borosilicate glasses.[91]

3.3.5. *Surface requirements for wafer direct bonding*

The bond strength and uniformity obtainable by direct bonding is generally controlled by particle contamination, surface roughness/flatness, and surface hydrophilicity; these considerations will now be examined in further detail. Particle contamination poses the greatest concern for most applications, because even relatively small micron-sized particles can create permanent cavities, called either voids or bubbles, several millimeters in size between the bonded wafers. As an alternative to commonly employed measures used to reduce particle counts in cleanroom and glovebox environments, wafers can be transferred directly from the final cleaning solution to a clean water stream where the initial spontaneous bond is performed. The second main requirement for direct bonding is sufficiently low surface roughness and surface flatness. Roughness is the local variation in wafer surface topology, whereas the maximum difference in surface height across the entire wafer is referred to as flatness (for a clamped wafer, measured with respect to the clamping surface) or bow (for an unclamped wafer). These are generally not a concern for most commercially available substrates, which have not undergone any processing other than cleaning. For processed substrates, as in the case of M/NEMS devices for which micro- or nanoscale channels or cavities must be sealed by the bonding process, the masking and etching processes involved in producing these structures can result in additional surface asperities and wafer bow. In this case, characterization of the surfaces may be necessary to determine if they are suitable for bonding. The effect of surface morphology on adhesion has been studied both theoretically and experimentally.[92–94,96] In their theoretical treatment of the subject, Gösele and coworkers describe two regimes: one in which the lateral extension of a gap, R, is much greater than the wafer thickness, t_w, and a second regime where the wafer thickness dominates.[95] In the first regime, a gap with height h will be closed provided it is less than a threshold value given by $h = R^2\sqrt{2Et_w^3/3y}$, while the threshold in the second regime is given by $h = 3.6(R\gamma/E)^{1/2}$. As a general guideline, Schmidt proposes that less than 10 Å roughness and less than 5 μm bow on a 4″ (102.6 mm) silicon wafer are required for complete direct bonding.[75] Christiansen and coworkers suggest a more conservative limit of 5 Å, based on the thickness of a double monolayer of physisorbed water, and report that wafer bow of up to 25 μm can be accommodated by elastic deformation for typical 0.5 mm thick 4″ silicon wafers.[71] The third main requirement for successful direct bonding, surface cleanliness, is important both from the point of view of maximizing the number of surface silanol groups available for reaction and minimizing the presence of organic contaminants, which can act as nucleation sites for agglomeration of bubble-producing gases.[96] For M/NEMS applications involving direct bonding of wafers with silicon dioxide surfaces, elimination of organic and particulate contaminants is typically achieved by cleaning the wafers in a piranha solution (H_2SO_4:H_2O_2 = 3:1 at room temperature) for 10–20 minutes, followed by rinsing in deionized water. Other common chemical surface treatments involving strong bases, such as concentrated hydrofluoric acid or the SC1 solution used in standard rolling circle amplification (RCA) cleaning (NH_4OH:H_2O_2:H_2O = 1:1:5, 75–80 °C), may not be appropriate for glass surfaces with nanoscale structure because of the solubility of glasses under these conditions. Surface hydrophylicity can also be increased by activating the surfaces with various plasma treatments, which is the subject of the following sub-section.

3.3.6. *Low temperature direct bonding by surface plasma activation*

A model for the molecular mechanisms involved in wafer direct bonding of silicon dioxide surfaces was first proposed by Stengl *et al.*,[89] and subsequently developed by Tong *et al.*[85,96] In this model, bonding occurs in two steps, the first being a restructuring of the three-dimensional hydrogen-bonded network of water molecules associated with opposing silanol groups, and the second being the formation of siloxane bridges, accompanied by the diffusive elimination of water. Kissinger and Kissinger showed experimentally that the bond energy kinetics for silicon wafers with a 500 nm wet thermal oxide layer annealed at 200 °C are consistent with an irreversible first-order reaction with an activation energy of 0.05 eV, a bonding site density of 4.1×10^{14} cm^{-2}, and a time constant of 8 hours.[90] Here, the activation energy is a theoretical value associated with the restructuring of the hydrogen bond network of water molecules, while the bonding site density and time constant are fitting parameters (the saturated silanol surface density is 4.6×10^{14} cm^{-2}). This result suggests that water restructuring is the dominant bonding mechanism at 200 °C, with no appreciable interfacial siloxane bridge formation, which is consistent with the earlier experimental result of Maszara *et al.*[79] On the other hand, the experiments of Tong and Gösele involving chemically oxidized silicon wafers, bearing only a native oxide layer 1–2 nm thick, indicated that water restructuring is only important for annealing temperatures up to 110 °C, and that interfacial siloxane bridge formation saturates at 150 °C.[85] Hence the thickness of the oxide layer, which affects the rate at which water is eliminated from the interface at a given temperature, is an important parameter for predicting the bonding reaction kinetics of a particular system.

Based on the above discussion, one would not expect any permanent covalent bonding of glass wafers to occur at temperatures below 400 °C. However, bond energies comparable to those achieved by high-temperature annealing at 1000 °C, or higher, have been demonstrated at temperatures as low as room temperature by treating the wafer surfaces with plasma prior to bonding.[90,98–104] Typically, oxygen plasma is generated using a commercial RIE system, either with or without ICP; although similar results have been obtained for argon-based RIE, and with nitrogen radical microwave plasmas. In the case of oxygen ICP-RIE treated silicon wafers, multiple internal reflection Fourier transform infrared spectra collected after treatment indicate an increase in the number of silanol groups compared to SC1 treated wafers.[98] However, the fact that bond strengthening can be accomplished with such a wide variety of plasma species has led to the hypothesis that the mechanism cannot be attributed entirely to a specific chemical reaction, but rather has its origins in the creation of interfacial porosity, leading to an increased rate of interfacial water removal.[99,100,103] It has also been observed that the surface energy exhibits a square-root dependence on anneal time for oxygen ICP treated silicon wafers, which supports a water diffusion related mechanism;[99] this result has not yet been validated for wafers with thick oxides or bulk glass materials.

3.3.7. *Anodic bonding*

Anodic bonding, which is also referred to as electrostatic bonding or field-assisted bonding, is a method for bonding metal or semiconductor

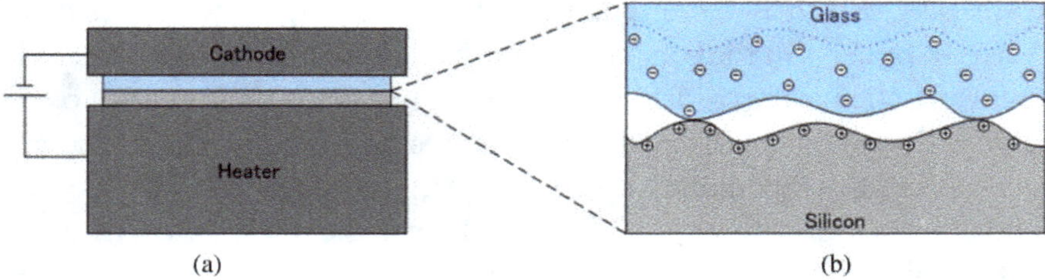

Figure 3.13. (a) A typical experimental setup of anodic wafer bonding, (b) Schematic illustration of the bonding mechanism.

wafers to glass wafers possessing sufficient ionic conductivity. The method was originally developed by Wallis and Pomerantz[105] and was recently reviewed in detail by Knowles and van Helvoort.[72] A typical experimental setup is shown in Figure 3.13(a), with an accompanying illustration of the bonding mechanism in Figure 3.13(b). Heating and applied potential are the basic requirements for all anodic bonding processes; commercial anodic bonding tools (e.g. Electronic Visions, Karl Suss) allow the user to control the temperature, potential, and external static force applied to the wafer stack, as well as the ambient pressure. Although this method has been applied to a variety of material combinations, silicon and Corning 7740 glass (Pyrex) are most commonly used, in part because their thermal strain difference is zero at a temperature of ~300 °C, which is well below the strain point of the glass (515 °C), and also due to the relatively low volume resistivity of Pyrex.[106,107] A plate cathode is advantageous in terms of electric field and temperature uniformity and stability, but in cases where bonding is carried out at atmospheric pressure, it can potentially trap gas pockets between the wafers due to multiple initial contact points. This problem can be alleviated by carrying out the bonding under vacuum conditions, or by employing a point cathode. Bonding in vacuum may also be advantageous for applications where one or both wafers contain microscale cavities, which can result in potentially damaging electrical discharge if filled with gases. The wafers are heated to a temperature which is ideally well below the strain point temperature of the glass material but high enough to significantly increase its ionic conductivity (typically 300–400 °C in the case of Pyrex). Once both wafers have reached the bonding temperature, a voltage typically in the range of 100–1500 V is applied to the wafer stack. As a result, ions in the glass migrate away from the bond interface, leaving behind a negative space charge layer, which is electrostatically attracted to the positively charged silicon anode and pulls the two surfaces into intimate contact. This makes it possible to achieve bonding at much lower temperatures than direct bonding techniques, which is one of its main advantages. Another benefit which is inherent in the anodic bonding mechanism is that the technique is very tolerant of surface roughness. It has been shown, theoretically, that the surface roughness must be larger than 1 μm before the electrostatic pressure is reduced by an order of magnitude.[108] By comparison, direct bonding processes can typically only tolerate roughness of 1 nm or less. The time required for complete anodic bonding, typically ~10 minutes, is also much faster than direct bonding methods that require far longer temperature ramp times.[109] Anodic bonds between Pyrex and silicon wafers, with tensile strengths of 10–15 MPa and hermetic wafer level seals, are routinely achieved.[107,110]

The mechanism for the creation of the space charge layer and subsequent intimate contact is currently understood only in qualitative terms.[108,111,112] The tensile strength of anodically bonded wafer stacks is maintained upon reversal of the applied potential, indicating that bonding is produced by one or more irreversible chemical reactions. It is generally believed that bonding results from the formation of siloxane bonds across the bond interface, either as a result of anodic oxidation of silicon or thermodehydration of silanol groups present on the anode (the latter reaction is discussed in further detail in Section 3.3.4). The oxidation of silicon is enhanced by an electric field.[113] Hence the focusing of the electric field on the depleted space charge region should be expected to give rise to an increased flux and reaction of oxygen at the anode. One model for the polarization of alkali glasses proposes that the conductivity is supported primarily by the hopping of monovalent cations (the most mobile being sodium ions in commercial glasses) between non-bonding oxygen (NBO) sites.[114,115] In this model, the interface between the anode material and the glass is treated as a blocking electrode, and conduction of sodium ions through the glass results in a negatively charged layer of unoccupied NBO sites, as well as accumulation of sodium ions at the cathode. At 475 °C, the mobility of sodium ions should be larger than that of NBO ions by a factor of ~10^7, so the latter are expected to provide most of the space charge.[115] Based on conservation of charge, the final depletion width obtained after bonding can be predicted by integrating the current time course measured during bonding experiments. However, when these predictions are compared to depletion widths measured directly by TEM, as in the experiments of van Helvoort and coworkers, the assumption that all of the measured charge originates from a single mobile cation species only accurately predicts the depletion width for the early stages of the bonding.[116] The authors found that a silicon-Pyrex bond carried out at 1 kV and 350 °C is consistent with depletion of a single positive charge carrier for approximately the first minute of bonding, after which it drastically over-predicts the observed depletion width. While this model is only quantitative on a short time scale, the depletion width experimentally achieves nearly three-quarters of its maximum value on this same time scale under the conditions described above. Thus the formation of the depletion region appears to be dominated by the conduction of a single positive charge carrier, with the persistence of current resulting from other conduction mechanisms. For example, a typical applied potential of 1 kV and a typical depletion width of 1 μm would produce an equilibrium electric field of 10^7 V cm^{-1}, similar to the dielectric breakdown electric field of silicon dioxide.[117] Hence, it is possible that electronic conduction could account for the persistence of current long after the depletion width has reached its equilibrium value.[118] An alternative explanation comes from nuclear magnetic resonance (NMR) experiments in which Schott 8330 (Tempax) glass was polarized via evaporated aluminum electrodes.[119] This data suggests that space charge is not due to NBO groups, but rather to BO_4^- groups, and that oxidation of the anode proceeds via transport of hydroxide molecules produced by ionization of native water molecules in the space charge layer. Yet another possibility is the removal of NBO ions from the glass, which was proposed as an explanation for observed local microporosity in polarized alkali glasses, resulting in sintering and increased density in a manner dependent on composition, temperature, and electric field.[120]

These and other effects are areas of active research in the development of more complete theoretical descriptions of the anodic bonding processes involving alkali glasses.

References

1. Abgrall P. and Nguyen N.T. (2008), Nanofluidics devices and their applications, *Anal Chem*, **80**, 2326–2341.
2. Mijatovic D., Eijkel J.C.T., and van den Berg A. (2005), Technologies for nanofluidic systems: *top-down* vs. *bottom-up* — a review, *Lab Chip*, **5**, 492–500.
3. Haneveld J., Jansen H., Berenschot E., Tas N., and Elwenspoek M. (2003), Wet anisotropic etching for fluidics 1D nanochannels, *J Micromech Microeng*, **13**, S62–S66.
4. Janssen K.G.H., Hoang H.T., Floris J., de Vries J., Tas N.R., Eijkel J.C.T., and Hankemeier T. (2008), Solution titration by wall deprotonation during capillary filling of silicon oxide nanochannels, *Anal Chem*, **50**, 8095–8101.
5. He Q., Chen S., Su Y., Fang Q., and Chen H. (2008), Fabrication of 1D nanofluidics channels on glass substrate by wet etching and room-temperature bonding, *Anal Chem Acta*, **628**, 1–8.
6. Mao P. and Han J. (2005), Fabrication and characterization of 20 nm planar nanofluidic channels by glass-glass and glass-silicon bonding, *Lab Chip*, **5**, 837–844.
7. Han J. and Craighead H.G. (1999), Entropic trapping and sieving of long DNA molecules in a nanofluidic channel, *J Vac Sci Technol A*, **17**, 2142–2147.
8. Chan Y.C., Zohar Y., and Lee Y.-K. (2009), Effects of embedded sub-micron pillar arrays in microfluidic channels on large DNA electrophoresis, *Electrophoresis*, **30**, 3242–3249.
9. Tsukahara T., Mawatari K., Hibara A., and Kitamori T. (2008), Development of a pressure-driven nanofluidic control system and its application to an enzymatic reaction, *Anal Bioanal Chem*, **391**, 2745–2752.
10. Hibara A., Takumi S., Kim H.-B., Tokeshi M., Ooi T., Nakao M., and Kitamori T. (2003), Nanochannels on a fused-silica microchip and

liquid properties investigation by time-resolved fluorescence measurements, *Anal Chem*, **36**, 605–612.

11. Tamaki E., Hibara A., Kim H.-B., Tokeshi M., and Kitamori T. (2006), Pressure-driven flow control system for nanofluidic chemical process, *J Chromatogr A*, **1137**, 256–262.
12. Renberg B., Sato K., Tsukahara T., Mawatari K., and Kitamori T. (2009), Hands on: thermal bonding of nano- and microfluidic chips, *Microchim Acta*, **166**, 177–181.
13. Stein D., Kruithof M., and Dekker C. (2004), Surface-charge-governed ion transport in nanofluidic channels, *Phys Rev Lett*, **93**, 035901-1–035901-4.
14. Kaji N., Tezuka Y., Takamura Y., Ueda M., Nishimoto T., Nakanichi H., Horiike Y., and Baba Y. (2004), Separation of long DNA molecules by quartz nanopillar chips under a direct current electric field, *Anal Chem*, **76**, 15–22.
15. Xia Q., Morton K.J., Austin R.H., and Chou S.Y. (2008), Sub-10 nm self-enclosed self-limited nanofluidic channel arrays, *Nano Lett*, **8**, 3830–3833.
16. O'Brien II M.J., Bisong P., Ista L.K., Rabinovich E.M., Garcia A.L., Sibbett S.S., Lopez G.P., and Brueck S.R.J. (2003), Fabrication of an integrated nanofluidic chip using interferometic lithography, *J Vac Sci Technol B*, **21**, 2941–2945.
17. Stavis S.M., Strychalski E.A., and Gaitan M. (2009), Nanofluidic structures with complex three-dimensional surfaces, *Nanotechnology*, **20**, 165302-1–165302-7.
18. Cao H., Tegenfeldt J.O., Austin R.H., and Chou S.Y. (2002), Gradient nanostructures for interfacing microfluidics and nanofluidics, *Appl Phys Lett*, **81**, 3058–3060.
19. Campbell L.C., Wilkinson M.J., Manz A., Camilleri P., and Humphreys C.J. (2004), Electrophoretic manipulation of single DNA molecules in nanofabricated capillaries, *Lab Chip*, **4**, 225–229.
20. Wang Y.M., Tegenfeldt J.O., Reisner W., Riehn R., Guan X.-J., Golding I., Cox E.C., Sturm J., and Austin R.H. (2005), Single-molecule studies of repressor-DNA interactions show long-range interactions, *Proc Natl Acad Sci USA*, **102**, 9796–9801.

21. Wang K.-G., Yue S., Wang L., Jin A., Gu C., Wang P.-Y., Feng Y., Wang Y., and Niu H. (2006), Manipulating DNA molecules in nanofluidic channels, *Microfluid Nanofluid*, **2**, 85–88.
22. Lee S., An R., and Hunt A.J. (2010), Liquid glass electrodes for nanofluidics, *Nature Nanotech*, **5**, 412–416.
23. Ke K., Hasselbrink, Jr. E.F., and Hunt A.J. (2005), Rapidly prototyped three-dimensional nanofluidic channel networks in glass substrates, *Anal Chem*, **77**, 5083–5088.
24. Han A., Mondin G., Hegelbach N.G., de Rooij N.F., and Staufer U. (2006), Filling kinetics of liquids in nanochannels as narrow as 27 nm by capillary force, *J Colloid Interf Sci*, **293**, 151–157.
25. Tas N.R., Mela P., Kramer T., Berenschot J.W., and van den Berg A. (2003), Capillarity induced negative pressure of water plugs in nanochannels, *Nano Lett*, **3**, 1537–1540.
26. Tas N.R., Berenschot J.W., Mela P., Jansen H.V., Elwenspoek M., and van den Berg A. (2002), 2D-confined nanochannels fabricated by conventional micromachining, *Nano Lett*, **2**, 1031–1032.
27. Schoch R.B., van Lintel H., and Renaud P. (2005), Effect of the surface charge on ion transport through nanoslits, *Phys Fluid*, **17**, 100604-1–100604-5.
28. Kamik R., Castelino K., Fan R., Yang P., and Majumdar A. (2005), Effects of biological reactions and modifications on conductance of nanofluidic channels, *Nano Lett*, **5**, 1638–1642.
29. Zevenbergen M.A.G., Krapf D., Zuiddam M.R., and Lemay S.G. (2007), Mesoscopic concentration fluctuations in a fluidic nanocavity detected by redox cycling, *Nano Lett*, **7**, 384–388.
30. Huang X.T., Gupta C., and Pennathur S. (2010), A novel fabrication method for centimeter-long surface-micromachined nanochannels, *J Micromech Microeng*, **20**, 015040-1–015040-9.
31. Mao P. and Han J. (2009), Massively-parallel ultra-high-aspect-ratio nanochannels as mesoporous membranes, *Lab Chip*, **9**, 586–591.
32. Eijkel J.C.T., Bomer J., Tas N.R., and van den Berg A. (2004), 1-D nanochannels fabricated in polyimide, *Lab Chip*, **4**, 161–163.
33. Hamblin M.N., Xuan J., Maynes D., Tolley H.D., Belnap D.M., Woolley A.T., Lee M.L., and Hawkins A.R. (2010), Selective trapping

and concentration of nanoparticles and viruses in dual-height nanofluidic channels, *Lab Chip*, **10**, 173–178.
34. Sordan R., Miranda A., Traversi F., Colombo D., Chrastina D., Isella G., Masserini M., Miglio L., Kern K., and Balasubamanian K. (2009), Vertical arrays of nanofluidic channels fabricated without nanolithography, *Lab Chip*, **9**, 1556–1560.
35. Zevenbergen M.A.G., Wolfrum B.L., Goluch E.D., Singh P.S., and Lemay S.G. (2009), Fast electro-transfer kinetics probed in nanofluidic channels, *J Am Chem Soc*, **131**, 11471–11477.
36. Goluch E.D., Wolfrum B., Singh P.S., Zevenbergen M.A.G., and Lemay S.G. (2009), Redox cycling in nanofluidic channels using interdigitated electrodes, *Anal Bioanal Chem*, **394**, 447–456.
37. Perssona F. and Tegenfeldt J.O. (2010), From microfluidic application to nanofluidic phenomena issue, *Chem Soc Rev*, **39**, 985–999.
38. Li W., Tegenfeldt J.O., Chen L., Austin R.H., Chou S.Y., Kohl P.A., Krotine J., and Sturm J.C. (2003), Sacrificial polymers for nanofluidic channels in biological applications, *Nanotechnology*, **14**, 548–583.
39. Jay Guo L., Cheng X., and Chou C.-F. (2004), Fabrication of size-controllable nanofluidic channels by nanoimprinting and its application for DNA stretching, *Nano Lett*, **4**, 69–73.
40. Zhang L., Gu F.X., Tong L.M., and Yin X.F. (2008), Simple and cost-effective fabrication of two-dimensional plastic nanochannels from silica nanowire templates, *Microfluid Nanofluid*, **6**, 727–732.
41. Kovarik M.L. and Jacobson S.C. (2007), Attoliter-scale dispensing in nanofluidic channels, *Anal Chem*, **79**, 1655–1660.
42. Liang X., Morton K.J., Austin R.H., and Chou S.Y. (2007), Single sub-20 nm wide, centimeter-long nanofluidic channel fabricated by novel nanoimprinting mold fabrication and direct imprinting, *Nano Lett*, **7**, 3774–3780.
43. Wang X., Liangjin G., Jingjing L., Xiaojun L., Keqiang Q., Yangchao T., Shaojun F., and Zheng C. (2009), Fabrication of enclosed nanofluidic channels by UV cured imprinting and optimized thermal bonding of SU-8 photoresist, *Microelectronic Eng*, **86**, 1347–1349.
44. Chung S., Lee J.H., Moon M.-W., Han J., and Kamm R.D. (2008), Non-lithographic wrinkle nanochannels for protein preconcentration, *Adv Mater*, **20**, 3011–3016.

45. Huh D., Mills K.L., Zhu X., Burns M.A., Thouless M.D., and Takayama S. (2007), Tuneable elastomeric nanochannels for nanofluidic manipulation, *Nat Mater*, 6, 424–428.
46. Efimenko K., Rackaitis M., Manias E., Vaziri A., Mahadevan L., and Genzer J. (2008), Nested self-similar wrinkling patterns in skins, *Nat Mater*, **20**, 3011–3016.
47. Chung J.Y., Nolte A.J., and Stafford C.M. (2009), Diffusion-controlled, self-organized growth of symmetric wrinkling patterns, *Adv Mater*, **21**, 1358–1362.
48. Xu B.-Y., Xu J.-J., Xia X.-H., and Chen H.-Y. (2010), Large scale lithography-free nano channel array on polystylene, *Lab Chip*, **10**, 2894–2901.
49. Sivanesan P., Okamoto K., English D., Lee C.S., and DeVoe D.L. (2005), Polymer nanochannels fabricated by thermomechanical deformation for single-molecule analysis, *Anal Chem*, **77**, 2252–2258.
50. Kim S.J. and Han J. (2008), Self-sealed vertical polymeric nanoporous-junctions for high-throughput nanofluidic applications, *Anal Chem*, **80**, 3507–3511.
51. Lee J.H., Song Y.-A., Tannenbaum S.R., and Han J. (2008), Increase of reaction rate and sensitivity of low-abundance enzyme assay using micro/nanofluidic preconcentration chip, *Anal Chem*, **80**, 3198–3204.
52. Flachbart B.R., Wong K., Iannacone J.M., Abante E.N., Vlach R.L., Rauchfuss P.A., Bohn P.W., Sweedler J.V., and Shannon M.A. (2006), Design and fabrication of a multilayered polymer microfluidic chip with nanofluidic interconnects *via* contact printing, *Lab Chip*, **6**, 667–674.
53. Fa K., Tulock J.J., Sweedler J.V., and Bohn P.W. (2005), Profiling pH gradient across nanocapillary array membranes connecting microfluidic channels, *J Am Chem Soc*, **127**, 13928–13933.
54. Liang X. and Chou S.Y. (2008), Nanogap detector inside nanofluidic channel for fast real-time label-free DNA analysis, *Nano Lett*, **8**, 1472–1476.
55. Goto M., Tsukahara T., Sato K., and Kitamori T. (2008), Micro- and nanometer scale patterned surface in a microchannel for cell

culture in microfluidic devices, *Anal Bioanal Chem*, **390**, 817–823.

56. Yamaguchi M., Nishimura O., Lim S., Shimokawa K., Tamura T., and Suzuki M. (2006), Protein patterning using a microstructured organosilane layer fabricated by VUV light lithography as a template, *Colloid Surface A: Physicochem Eng Aspects*, **284–285**, 532–534.
57. Gan J., Chen H., Zhou F., Huang H., Zheng J., Song W., Yuan L., and Wu Z. (2010), Fabrication of cell pattern on poly(dimethylsiloxane) by vacuum ultraviolet lithography, *Colloid Surface B: Biointerfaces*, **76**, 381–385.
58. Yap F.L. and Zhang T. (2007), Protein and cell micropatterning and its integration with micro/nanoparticles assembly, *Biosens Bioelectron*, **22**, 775–788.
59. Falconnet D., Csucs G., Grandin H.M., and Textor M. (2006), Surface engineering approaches to micropattern surfaces for cell-based assays, *Biomaterials*, **27**, 3044–3063.
60. Idota N., Tsukahara T., Sato K., Okano T., and Kitamori T. (2009), The use of electron beam lithographic graft-polymerization on thermoresponsive polymers for regulating the directionality of cell attachment and detachment, *Biomaterials*, **30**, 2095–2100.
61. Jang K., Sato K., Tanaka Y., Xu Y., Sato M., Nakajima T., Mawatari K., Konno T., Ishihara K., and Kitamori T. (2010), An efficient surface modification using 2-methacryloyloxyethyl phosphorylcholine to control cell attachment via photochemical reaction in a microchannel, *Lab Chip*, **10**, 1937–1945.
62. Renberg B., Sato K., Mawatari K., Idota N., Tsukahara T., and Kitamori T. (2009), Serial DNA immobilization in micro- and extended nanospace channels, *Lab Chip*, **9**, 1517–1523.
63. Park J.U., Lee J.H., Paik U., Lu Y., and Rogers J.A. (2008), Nanoscale patterns of oligonucleotides formed by electrohydrodynamic jet printing with applications in biosensing and nanomaterials assembly, *Nano Lett*, **8**, 4210–4216.
64. Mooney J.F., Hunt A.J., Mcintosh J.R., Liberko C.A., Walba D.M., and Rogers C.T. (1996), Patterning of functional antibodies and other proteins by photolithography of silane monolayers, *Proc Natl Acad Sci USA*, **93**, 12287–12291.

65. Jiang J., Li X., Mak W.C., and Trau D. (2008), Integrated direct DNA/protein patterning and microfabrication by focused ion beam milling, *Adv Mater*, **20**, 1636–1643.
66. Kane R.S., Takayama S., Ostuni E., Ingber D.E., and Whitesides G.M. (1999), Patterning proteins and cells using soft lithography, *Biomaterials*, **20**, 2363–2376.
67. Hoff J.D., Cheng L.J., Meyhöfer E.L., Guo J.L., and Hunt A.J. (2004), Nanoscale protein patterning by imprint lithography, *Nano Lett*, **4**, 853–857.
68. Cao H., Yu Z., Wang J., Tegenfeldt J.O., Austin R.H., Chen E., Wu W., and Chou S.Y. (2002), Fabrication of 10 nm enclosed nanofluidic channels, *Appl Phys Lett*, **81**, 174–176.
69. Tong Q.-Y. and Gösele U. (1999), *Semiconductor Wafer Bonding Science and Technology*, New York: John Wiley & Sons, Inc.
70. D'Aragona F.S. and Ristic L. (1994), 'Chapter 5' in *Sensor Technology and Devices*, Ristic L. (Ed.), Boston: Artech House.
71. Christiansen S.H., Singh R., and Gosele U. (2006), Wafer direct bonding: from advanced substrate engineering to future applications in micro/nanoelectronics, *Proc IEEE*, **94**, 2060–2106.
72. Knowles K.M. and van Helvoort A.T.J. (2006), Anodic bonding, *Int Mater Rev*, **51**, 273–311.
73. Reiche M. (2006), Semiconductor wafer bonding, *Phys Status Solidi A*, **203**, 747–759.
74. Gösele U. and Tong Q.Y. (1998), Semiconductor wafer bonding, *Ann Rev Mat Sci*, **28**, 215–241.
75. Schmidt M.A. (1998), Wafer-to-wafer bonding for microstructure formation, *Proc IEEE*, **86**, 1575–1585.
76. Lemons R.A. and Quate C.F. (1974), Acoustic microscope — scanning version, *Appl Phys Lett*, **24**, 163–165.
77. Lang A.R. (1958), Direct observation of individual dislocations X-ray diffraction, *J Appl Phys*, **29**, 597–598.
78. Bengtsson S. (1992), Semiconductor wafer bonding — a review of interfacial properties and applications, *J Electron Mater*, **21**, 841–862.
79. Maszara W.P., Goetz G., Caviglia A., and McKitterick J.B. (1988), Bonding of silicon-wafers for silicon-on-insulator, *J Appl Phys*, **64**, 4943–4950.

80. Gillis P.P. and Gilman J.J. (1964), Double-cantilever cleavage mode of crack propagation, *J Appl Phys*, **35**, 647–658.
81. Müller B. and Stoffel A. (1991), Tensile strength characterization of low-temperature fusion-bonded silicon wafers, *J Micromech Microeng*, **1**, 161–166.
82. Maleville C. and Mazure C. (2004), Smart-Cut (R) technology: from 300 mm ultrathin SOI production to advanced engineered substrates, *Solid State Elec*, **48**, 1055–1063.
83. Al-Sarawi S.F., Abbott D., and Franzon P.D. (1998), A review of 3-D packaging technology, *IEEE T Compon Pack B*, **21**, 2–14.
84. Esashi M. (2008), Wafer level packaging of MEMS, *J Micromech Microeng*, **18**, 1–13.
85. Tong Q.Y. and Gösele U. (1996), A model of low-temperature wafer bonding and its applications, *J Electrochem Soc*, **143**, 1773–1779.
86. Lasky J.B. (1986), Wafer bonding for silicon-on-insulator technologies, *Appl Phys Lett*, **48**, 78–80.
87. Grundner M. and Jacob H. (1986), Investigations on hydrophilic and hydrophobic silicon (100) wafer surfaces by x-ray photoelectron and high-resolution electron-energy loss-spectroscopy, *Appl Phys A-Mater*, **39**, 73–82.
88. Taft E.A. (1980), Index of refraction of steam grown oxides on silicon, *J Electrochem Soc*, **12**, 993–994.
89. Stengl R., Tan T., and Gösele U. (1989), A model for the silicon-wafer bonding process, *Jpn J Appl Phys 1*, **28**, 1735–1741.
90. Kissinger G. and Kissinger W. (1993), Void-free silicon-wafer-bond strengthening in the 200–400 °C range, *Sensor Actuat A-Phys*, **36**, 149–156.
91. Field L.A. and Muller R.S. (1990), Fusing silicon-wafers with low melting temperature glass, *Sensor Actuat A-Phys*, **23**, 935–938.
92. Turner K.T. and Spearing S.M. (2002), Modeling of direct wafer bonding: effect of wafer bow and etch patterns, *J Appl Phys*, **92**, 7658–7666.
93. Maugis D. (1996), On the contact and adhesion of rough surfaces, *J Adhes Sci Technol*, **10**, 161–175.
94. Maszara W.P., Jiang B.L., Yamada A., Rozgonyi G.A., Baumgart H., and de Kock A.J.R. (1991), Role of surface-morphology in wafer bonding, *J Appl Phys*, **69**, 257–260.

95. Tong Q.Y. and Gösele U. (1995), Thickness considerations in direct silicon-wafer bonding, *J Electrochem Soc*, **142**, 3975–3979.
96. Mitani K., Lehmann V., Stengl R., Feijoo D., Gosele U., and Massoud H.Z. (1991), Causes and prevention of temperature-dependent bubbles in silicon-wafer bonding, *Jpn J Appl Phys 1*, **30**, 615–622.
97. Tong Q.Y., Cha G., Gafiteanu R., and Gosele U. (1994), Low-temperature wafer direct bonding, *J Microelectromech S*, **3**, 29–35.
98. Amirfeiz P., Bengtsson S., Bergh M., Zanghellini E., and Börjesson L. (2000), Formation of silicon structures by plasma-activated wafer bonding, *J Electrochem Soc*, **147**, 2693–2698.
99. Bengtsson S. and Amirfeiz P. (2000), Room temperature wafer bonding of silicon, oxidized silicon, and crystalline quartz, *J Electron Mater*, **29**, 909–915.
100. Zucker O., Langheinrich W., Kulozik M., and Gooebel H. (1993), Application of oxygen plasma processing to silicon direct bonding, *Sensor Actuat A-Phys*, **36**, 227–231.
101. Farrens S.N., Dekker J.R., Smith J.K., and Roberds B.E. (1995), Chemical free room-temperature wafer to wafer direct bonding, *J Electrochem Soc*, **142**, 3949–3955.
102. Sanz-Velasco A., Amirfeiz P., Bengtsson S., and Colinge C. (2003), Room temperature wafer bonding using oxygen plasma treatment in reactive ion etchers with and without inductively coupled plasma, *J Electrochem Soc*, **150**, G155–G162.
103. Howlader M.M.R., Suehara S., and Suga T. (2006), Room temperature wafer level glass/glass bonding, *Sensor Actuat A-Phys*, **127**, 31–36.
104. Howlader M.M.R., Suehara A., Takagi H., Kim T.H., Maeda R., and Suga T. (2006), Room-temperature microfluidics packaging using sequential plasma activation process, *IEEE Trans Adv Pack*, **29**, 448–456.
105. Wallis G. and Pomerantz D.I. (1969), Field assisted glass-metal sealing, *J Appl Phys*, **40**, 3946–3949.
106. Bansal N.P. and Doremus R.H. (1986), 'Chapter 3' in *Handbook of Glass Properties*, Orlando, FL: Academic Press.
107. Rogers T. and Kowal J. (1995), Selection of glass, anodic bonding conditions and material compatibility for silicon-glass capacitive sensors, *Sensor Actuat A-Phys*, **46**, 113–120.

108. Anthony T.R. (1983), Anodic bonding of imperfect surfaces, *J Appl Phys*, **54**, 2419–2428.
109. Cozma A., Jakobsen H., and Puers R. (1998), Electrical characterization of anodically bonded wafers, *J Micromech Microeng*, **8**, 69–73.
110. Cozma A. and Puers B. (1995), Characterization of the electrostatic bonding of silicon and Pyrex glass, *J Micromech Microeng*, **5**, 98–102.
111. Albaugh K.B. (1991), Electrode phenomena during anodic bonding of silicon to sodium borosilicate glass, *J Electrochem Soc*, **138**, 3089–3094.
112. Kanda Y., Matsuda K., Murayama C., and Sugaya J. (1990), The mechanism of field-assisted silicon glass bonding, *Sensor Actuat A-Phys*, **23**, 939–943.
113. Jorgensen J. (1962), Effect of an electric field on silicon oxidation, *J Chem Phys*, **37**, 874–877.
114. Doremus R.H. (1994), 'Chapter 9' in *Glass Science*, 2nd Ed., Chichester: John Wiley & Sons, Inc.
115. Carlson D.E. (1974), Ion depletion of glass at a blocking anode 1. Theory and experimental results for alkali silicate-glasse, *J Am Ceram Soc*, 57, 291–294.
116. van Helvoort A.T.J., Knowles K.M., and Fernie J.A. (2003), Characterization of cation depletion in pyrex during electrostatic bonding, *J Electrochem Soc*, **150**, G624–G629.
117. Lynch W.T. (1972), Calculation of electric-field breakdown in quartz as determined by dielectric dispersion analysis, *J Appl Phys*, **43**, 3274–3278.
118. Krieger U.K. and Lanford W.A. (1988), Field assisted transport of Na^+ ions, Ca^{2+} ions and electrons in commercial soda-lime glass 1: Experimental, *J. Non-Cryst Solids*, **102**, 50–61.
119. Nitzsche P., Lange K., Schmidt B., Grigull S., and Kreissig U. (1998), Ion drift processes in pyrex-type alkali-borosilicate glass during anodic bonding, *J Electrochem Soc*, **145**, 1755–1762.
120. Carlson D.E., Hang K.W., and Stockdale G.F. (1974), Ion depletion of glass at a blocking anode: II. Properties of ion-depleted glasses, *J Am Ceram Soc*, **57**, 295–300.

Chapter 4

FUNDAMENTAL TECHNOLOGY: FLUIDIC CONTROL METHODS

Fluidic control is one of the basic technologies for integrated chemical systems, in addition to fabrication, detection, and surface modification. Recently, integrated systems have been downscaling to 10^{-8} – 10^{-6} m (extended nanoscale) dimensions, which bridges the gap between single molecules and condensed phases. A nanofluidic control method is required to develop novel applications utilizing unique fluid properties in this scale, such as higher viscosity, lower permittivity, and high proton mobility.[1,2] Currently, fluidic control by electroosmotic flow and electrophoresis, generated by an external electric field, is primarily used for extended nanofluidic systems due to easy operation.[3] This is partially because fluidic control by pressure-driven flow becomes difficult in extended nanospace, due to the high fluidic resistance. Nevertheless, pressure-driven flow can provide for general chemical systems, as was verified in microtechnologies, and is expected to be a key technology in fully exploiting nanofluidic properties. Shear-driven flow can be another solution in overcoming the problems of high fluidic resistance. This chapter covers basic fluidic theory and recent advances in fluidic control methods for extended nanospace.

4.1. Basic Theory

The flows in nanofluidic systems of 10^{-9} m scale are so small that the continuum assumption often breaks down. Flow regimes are associated with the continuum assumption of the Knudsen number, given by

$$Kn = \frac{\lambda}{L_c} \tag{1}$$

where λ is the molecular length scale characterizing the fluid structure, and L_c the characteristic size of the flow.[4] A classification of the different flow regimes is given as follows:

$Kn < 10^{-2}$	: continuum flow
$10^{-2} < Kn < 10^{-1}$	: slip flow
$10^{-1} < Kn < 3$	: transitional regime
$Kn > 3$	: free molecular flow

For gas flows, the classifications are due to the mean path length of a molecule of air, approximately 70 nm at 100 Pa and 25 °C. The Knudsen number of a gas flow in extended nanospace ranges from 0.07–7, which corresponds to the slip flow or transition regime. For liquid flows of condensed phase, the molecular length scale is determined by the order of the molecular diameter.[5] In the case of water, with a molecular diameter of 3 Å, the continuum regime holds down to 10^{-8} m. Therefore, the gas and liquid flows have different regimes in extended nanospace.

In extended nanospace, the liquid flow is in the continuum regime and nearly incompressible (the fluid density, ρ, is constant). Liquids such as water are regarded as Newtonian fluids, where the shear stress is proportional to the strain rate (the velocity gradient).[4] Hence, the equation of continuity and the momentum equation, i.e. the Navier–Stokes equation, is given by

$$\nabla \cdot \mathbf{u} = 0. \tag{2}$$

$$\rho\left(\frac{\partial \mathbf{u}}{\partial t} + \mathbf{u} \cdot \nabla \mathbf{u}\right) = -\nabla p + \mathbf{f}_b + \mu \nabla^2 \mathbf{u}, \tag{3}$$

where $\mathbf{u}$ is the fluid velocity, t is the time, p is the pressure, $\mathbf{f}_b$ is the external body force, and μ is the fluid viscosity. By using dimensionless variables, $\mathbf{x}^* = \mathbf{x}/L_c$, $t^* = t/\tau_c$, $\mathbf{u}^* = \mathbf{u}/U_c$, and $p^* = pL_c/\mu U_c$, where τ_c is the characteristic time and U_c is the characteristic velocity, equation (3) is normalized as follows

$$Re\left(\frac{1}{Sr}\frac{\partial \mathbf{u}^*}{\partial t^*} + \mathbf{u}^* \cdot \nabla^* \mathbf{u}^*\right) = -\nabla^* p^* + \mathbf{f}_b{}^* + \nabla^{*2} \mathbf{u}^*, \tag{4}$$

where $\mathbf{f_b}^* = (L_c^2/\mu U_c)\mathbf{f_b}$ is the dimensionless body force. Here, two important dimensionless numbers are derived. One is the Reynolds number *Re*, i.e. the ratio of the inertial to the viscous forces,

$$Re = \frac{\rho U_c L_c}{\mu}. \tag{5}$$

Another is the Strouhal number, i.e. the ratio of the characteristic time of the unsteady force to the convective time scale,

$$Sr = \frac{\tau_c U_c}{L_c}. \tag{6}$$

For extended nanospace channel flows, assume that $L_c = 10^{-7}$ m and $U_c = 10^{-3}$ m at a viscosity of 10^{-3} Pa s and a density of 10^3 kg m^{-3}. In this case, $Re = 10^{-4}$, meaning the inertial force is negligible compared to the viscous force, i.e. the laminar flow regime. When $Re/Sr << 1$ ($\tau_c > 10^{-8}$ s) or the flow is in the steady state, equation (4) reduces to the Stokes equation,

$$0 = -\nabla^* p^* + \mathbf{f_b}^* + \nabla^{*2}\mathbf{u}^*. \tag{7}$$

This equation indicates that the fluid can be driven by the pressure, external body force, and viscous force.

The boundary conditions for the fluidic systems are one of the most important components to determine the flows. Generally, the fluid dynamics are based upon the no-slip boundary condition, i.e. the liquid next to a solid surface moves with the same velocity as that of the surface. The no-slip boundary condition is considered to be due to the trapping of liquid in pockets and crevices on the solid, and the attractive forces between the molecules of the solid and the liquid.[6] On the other hand, slippage over a non-wetting (i.e. hydrophobic, for the case of water) surface, even at microscales not in the slip flow regime, has been discovered from both experimental and numerical studies.[6–8] The slip produces friction drag reduction and affects mass, momentum, and energy

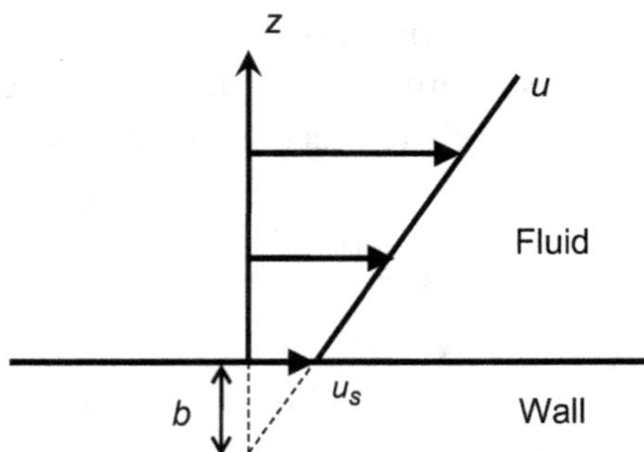

Figure 4.1. Schematic illustration of the slip at the wall.

transport at the fluid-solid interface. The superhydrophobic surface generated by micro- or nanostructures on a channel wall surface can enhance the slippage of liquid because of the entrapped gas in these structures. The magnitude of slippage is described by using the slip length b, related to the liquid velocity at the wall, u_s, as defined in Figure 4.1,

$$u_s = b\left.\frac{\partial u}{\partial z}\right|_{z=0}. \tag{8}$$

Experimental work has reported slip lengths in the range of –5–250 μm, which are dependent on the liquid, solid material, and measurement method.[7]

In reality, fluidic systems have various geometries of channel. The characteristic length of these geometries is represented by the hydraulic diameter, D_h, given by

$$D_h = \frac{4A}{L_p}, \tag{9}$$

where A is the area of the channel and L_p is the wetted perimeter. By using this parameter, the flows in the system can be regarded as the flow in a cylindrical channel with a diameter of D_h. However, it is noted that there are, of course, differences between a solution derived from equation (9) with the cylindrical channel model, and that directly solved considering the original channel geometry.[9]

The small scale flows are affected by the surface electric potential with the electric double layer, i.e. a layer of accumulated counter-ions near the wall. The electroneutrality of fluid breaks down in this region and the electric charge density, ρ_E, exerts a body force given by

$$\mathbf{f}_b = \rho_E \mathbf{E}, \tag{10}$$

where **E** is the electric field. The electric charge density related to the ion distribution is governed by the Poisson–Boltzmann equation as follows[10]

$$\nabla^2 \psi = -\frac{\rho_E}{\varepsilon} = -\frac{F}{\varepsilon}\sum z_i c_{0,i} \exp\left(-\frac{z_i F \psi}{RT}\right), \tag{11}$$

where ψ is the electric potential, F is the Faraday constant, ε is the electrical permittivity, z_i is the ion valence of ith species, $c_{0,i}$ is the bulk concentration, R is the gas constant, and T is the temperature. The length scale of the electrically polarized region, i.e. the thickness of the electric double layer, is defined by the Debye length κ^{-1}

$$\kappa^{-1} = \left(\frac{\varepsilon RT}{F^2 \sum z_i^2 c_{0,i}}\right)^{\frac{1}{2}}. \tag{12}$$

For a monovalent and symmetric electrolyte, the Debye length is calculated to be $\kappa^{-1} = 1$ μm at $c_{0,i} = 10^{-7}$ M, and $\kappa^{-1} = 10$ nm at $c_{0,i} = 10^{-3}$ M. An important parameter to characterize the flow is the ratio of the channel scale to the Debye length, κL_c. Most microfluidic systems are based upon the thin electric double layer approximation at $\kappa L_c >> 1$, and the theory of electrokinetics (i.e. electroosmosis, electrophroesis, and streaming potential/current) is well established. On the other hand, in extended nanospace, electric double layer overlap, with $\kappa L_c < 1$, often occurs for dilute electrolyte solutions. In this regime, phenomena are not well explored and a lot of research is ongoing.[11–13]

In addition to the above factors, there is a possibility that the liquid properties in extended nanospace are different from the bulk properties. Kitamori and coworkers highlighted that the viscosity of water can also be affected by H_2O–SiOH interactions on a glass surface confined in 2-D extended nanospace.[14]

4.2. Pressure-Driven Flow

Pressure-driven flow is broadly utilized in microfluidic systems. The flow is generated by the pressure gradient across the channel. It is appropriate for general chemical operations including solvent extraction and reaction. The velocity profile in the steady state is solved from equation (7) at no-slip boundary conditions ($\mathbf{u} = 0$ at the walls). When the electric body force is negligible (i.e. an electroneutral fluid with thin electric double layer, $\kappa L_c >> 1$), the flow has a parabolic velocity profile. The velocity of the flow in the cylindrical channel of hydraulic diameter D_h (Figure 4.2(a)) is given by

$$u = -\frac{D_h^2}{16\mu}\frac{\partial p}{\partial x}\left(1 - \frac{4r^2}{D_h^2}\right). \tag{13}$$

(a)

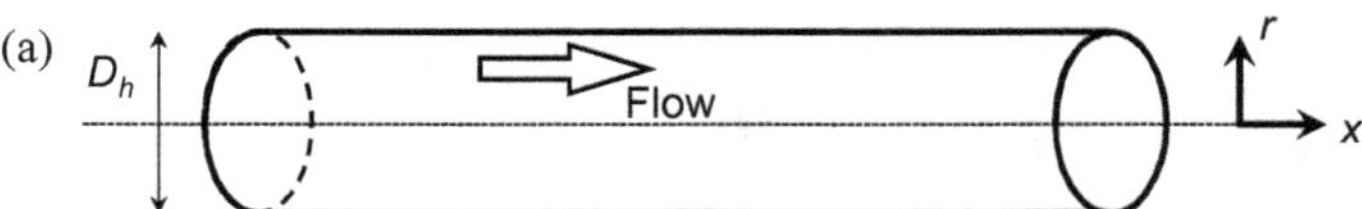

(b)

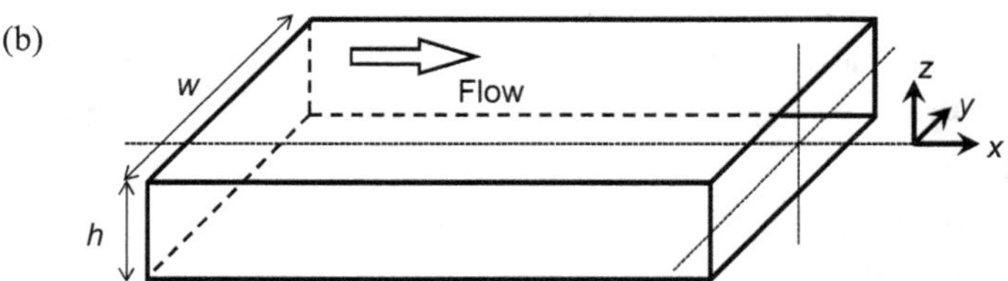

Figure 4.2. Geometry and the coordinate system of (a) cylindrical and (b) rectangular channel.

From the integral of equation (13) and $\partial p/\partial x = \Delta p/L$, where L is the channel length, the pressure drop in the channel is obtained

$$\Delta p = \frac{128\mu L}{\pi {D_h}^4} Q, \tag{14}$$

where Q is the volumetric flow rate. This equation is well known as the Hagen–Poiseuille law, which shows the friction drag in laminar flows. On the other hand, the velocity profile in a rectangular channel of width w and height h (Figure 4.2(b)) is obtained as follows[15,16]

$$u = \left(-\frac{h^2}{8\mu}\frac{\partial p}{\partial x}\right)\frac{32}{\pi^3}\sum_{i=0}^{\infty}\frac{(-1)^i}{(2i+1)^3}\left\{1-\frac{\cosh\left[\frac{(2i+1)\pi y}{h}\right]}{\cosh\left[\frac{(2i+1)\pi w}{2h}\right]}\right\} \cos\left[\frac{(2i+1)\pi z}{h}\right]. \tag{15}$$

For a thick electric double layer or double layer overlap, i.e. $\kappa L_c < 10$, the fluid in the channel is no longer electroneutral and the electric body force should be taken into account. When the fluid is driven in the channel, the flow leads to an ion flow of the electric double layer. This ion flow gives rise to the streaming potential/current across the channel (detailed in a later section). The induced electric field in turn exerts the electric body force on the fluid, termed the electroviscosity effect.[17,18] Ren and Li simulated the ion distribution in the channel at a thick electric double layer, based on the Nernst equation, and calculated a velocity profile for the pressure-driven flow. The results indicated that the velocity profile is different from the parabolic profile, and the volumetric flow rate lowers.[17]

For the extended nanospace channels, fluidic control systems providing both a high pressure and low flow rate are required due to its high pressure drop. If the fluid is driven at a velocity of 1 mm s^{-1} in an extended nanospace channel with a 100 nm hydraulic diameter, the pressure and volumetric flow rate are 1 MPa and 1 pL min^{-1},

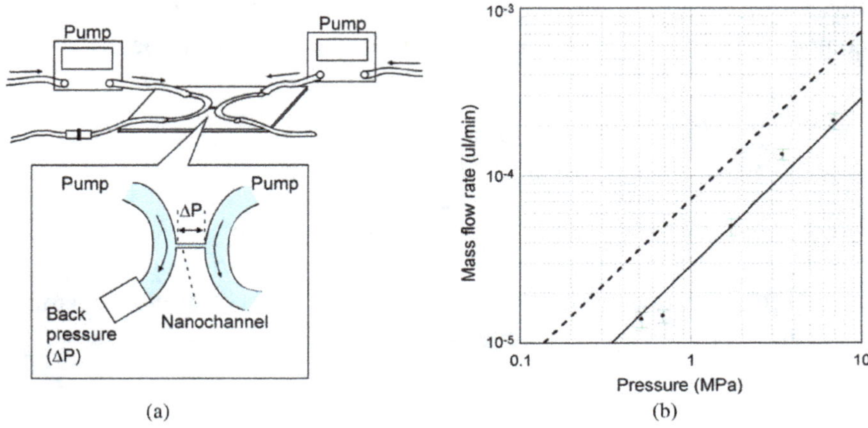

Figure 4.3. (a) Schematic illustration of a backpressure regulation system, (b) The mass flow rate in the nanochannels (380 nm width, 240 nm depth and 190 μm length, 40 channels) as a function of the applied pressure. The broken line plots Hagen–Poiseuille low at the viscosity of 0.001 Pa s, while the solid line at the viscosity of 0.0025 Pa s (Adapted from Ref. 19).

respectively. Namely, commercial syringe pumps cannot be used to drive a liquid in extended nanospace. Therefore, Kitamori and coworkers developed a backpressure-based fluidic control system, comprised of a backpressure regulator and high performance liquid chromatography (HPLC) pumps, as illustrated in Figure 4.3(a).[19] After an aqueous solution containing resorcinol probe molecules was introduced into the U-shaped, left-side microchannel using a HPLC pump at ~MPa, the solution could proceed inside the extended nanospace channels. The solution was recovered after passing through the extended nanospace channels, with another aqueous solution, including methylresorcinol as an internal standard molecule, in a U-shaped right-side microchannel. The volumetric flow rate in the extended nanospace channels was measured from the ratio of the probe molecule to the internal standard molecule (Figure 4.3(b)). The results showed that the obtained flow rate depends linearly on the applied pressure, and is lower than that predicted by equation (13). This difference between the experimental and theoretical values could be because of the 2~3 times larger viscosity in the extended nanospace

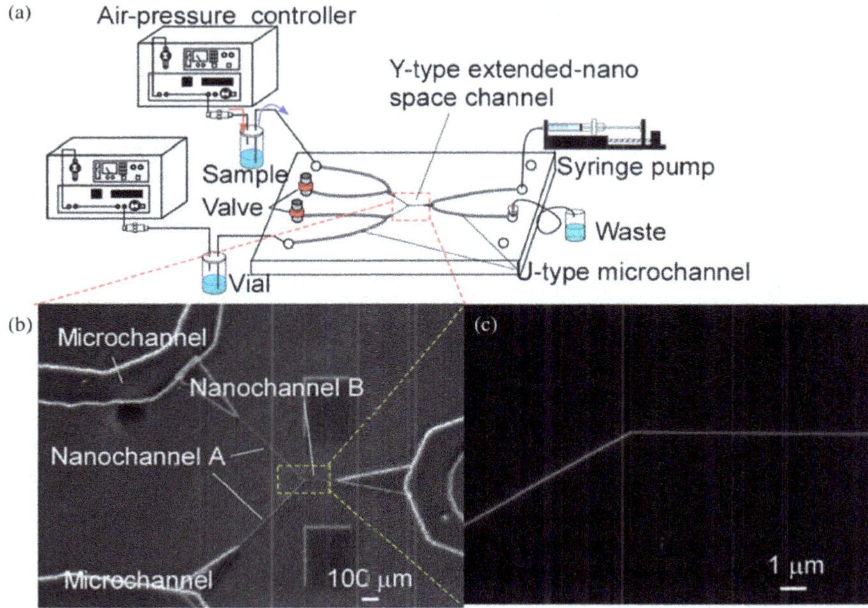

Figure 4.4. (a) Schematic illustration of the air-pressure-based nanofluidic control system, (b) Enlarged view of the fabricated microchip with the U-shaped microchannels and the Y-shaped extended nanospace channel, (c) A fluorescence image of the mixing of a fluorescein solution and a buffer solution flowing through the nanochannel (Adapted from Ref. 20).

compared with the bulk. However, since this method depended on the response time of backpressure regulators, the chaotic flow conditions, such as backward flow or plug-like flow, took place in the nanochannel. Kitamori *et al.* improved the system to an air-pressure-based nanofluidic control system, as shown in Figure 4.4(a), and evaluated its performance.[20] Liquid introduction and mixing of different solutions was achieved precisely, by applying the air-pressure to a Y-shaped extended nanospace channel through U-shaped microchannels on a glass microchip (Figures 4.4(b) and (c)). The performance of the system was evaluated by detecting the fluorescent intensities before and after the mixing of a fluorescein solution and buffer solution. The flow could be controlled in the pressure range of 0.003–0.4 MPa and flow rate range of 0.16–21.2 pL min^{-1}.

4.3. Shear-Driven Flow

Shear-driven flow, the so-called Couette flow, is expected to be an alternative fluidic control method to pressure-driven flow. The flow is induced by moving a surface toward a fixed plate, as illustrated in Figure 4.5. Different materials, such as a flat plate, rotation disk, or flexible belt, can be used as the moving surface. Since the fluid is driven in the gap between the plates according to the viscous force, it is possible to control the fluid velocity through the surface's speed of movement, even in the extended nanospace.

Consider an infinite moving plate with a constant velocity u_w parallel to a fixed plate at $\nabla p = 0$, and with a negligible electric body force (i.e. an electroneutral fluid with thin electric double layer, $\kappa L_c >> 1$). The velocity profile is obtained by solving equation (7) with boundary conditions of $u = 0$ at $z = -h/2$, and $u = u_w$ at $z = h/2$

$$u = u_w \left(\frac{1}{2} + \frac{z}{h} \right). \tag{16}$$

This equation indicates that the shear-driven flow has a linear velocity profile. However, in an extended nanospace with thick electric double layer $\kappa L_c < 10$, the electroviscosity effect can affect the flow profile, as well as the pressure-driven flow.

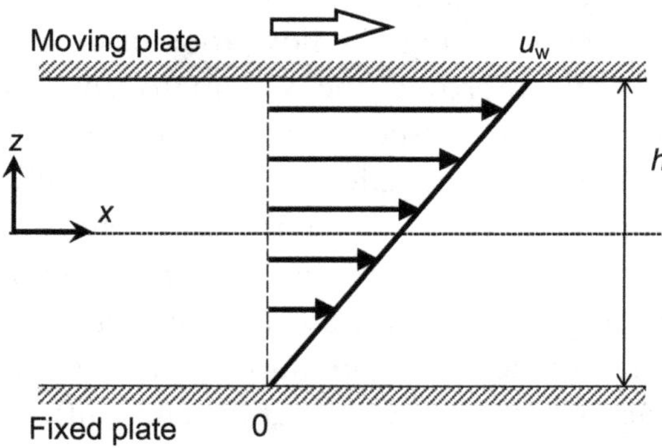

Figure 4.5. Schematic illustration of the shear-driven flow.

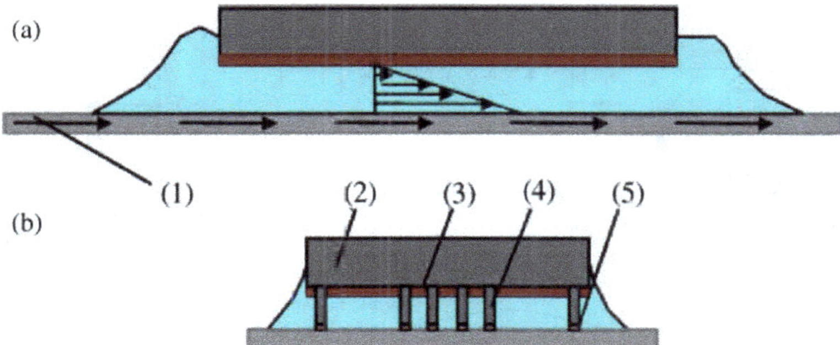

Figure 4.6. (a) Schematic cross section and (b) longitudinal view of the shear-driven flow system, showing the dragging effect of the moving channel wall (1) and the resulting linear velocity profile. The etched fused-silica platelet, the 10 mm × 20 mm stationary channel wall (2) put on the top of this wafer, carried the C18 layer (3) and contained an array of parallel spacers (4). The lubrication layer (5) between the spacers and the moving wall is also illustrated (Adopted from Ref. 23).

Desmet *et al.* has developed nanofluidic control systems based upon the shear-driven flows.[21–23] An example of the shear-driven flow system is illustrated in Figure 4.6. The 1-D nanochannels were assembled by pressing two flat substrates against each other. The upper plate is a rectangular fused-silica platelet, with an array of parallel channel spacers on its surface. The lower plate is a fully flat fused-silica wafer with a flatness of $\lambda/10$, moved with a displacement stage. The mean height of the spacers was 300 nm, which were fabricated by etching the upper glass plate. After the solution was introduced into the nm-sized gap between the plates by capillary action, the fluid inside the gap was driven by moving the lower plate. The velocity could be controlled in the range of 1–35 mm s^{-1}.

4.4. Electrokinetically-Driven Flow

Electrokinetically-driven flow is the flow generated by an external electric field, as illustrated in Figure 4.7. The fluid flow is generated by the electric body force. Hence the flow is free from the fluidic resistance, and suitable as a driving force for micro- and nanofluidic

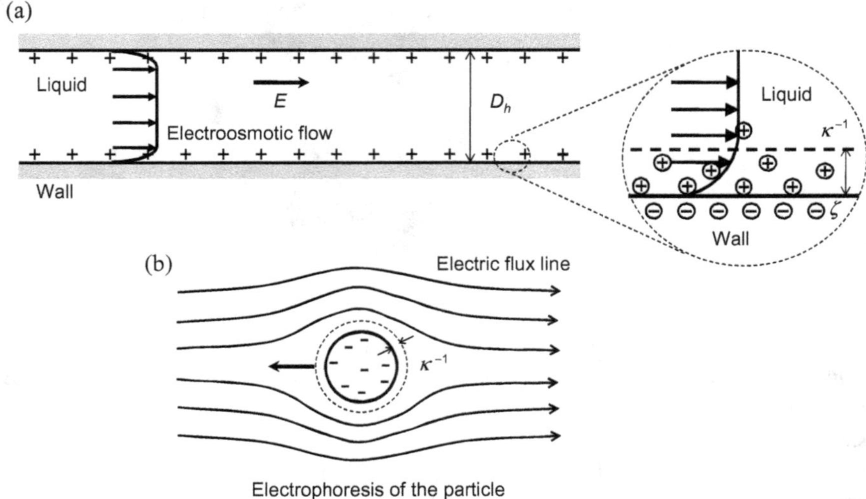

Figure 4.7. Schematic illustration of electrokinetically-driven flow: (a) Electrosomotic flow of the liquid, (b) Electrophoresis of the particle.

systems. In addition, molecules and particles which are suspended in the liquid move depending on their electric charge. Therefore, the sample separation, called capillary electrophoresis, can be easily realized for integrated chemical systems.

When a DC electric field E is applied across the channel, the fluid flow and particle migration occur simultaneously, electrosomotic flow and electrophoresis, respectively. Therefore, the particle velocity is a superposition of the electroosmotic velocity, $\mathbf{u}_{\mathbf{EOF}}$, and the electrophoretic velocity, $\mathbf{u}_{\mathbf{EP}}$, as follows

$$\mathbf{u}_{\mathbf{p}} = \mathbf{u}_{\mathbf{EOF}} + \mathbf{u}_{\mathbf{EP}} = \mu_{EOF}\mathbf{E} + \mu_{EP}\mathbf{E}. \tag{17}$$

These velocities are proportional to the electric field. The magnitude of the velocity per unit electric field is represented as the electroosmotic mobility, μ_{EOF}, and the electrophoretic mobility, μ_{EP}.

Electroosmotic flow is induced by the electric body force, with the migration of diffused ions in the electric double layer as illustrated in Figure 4.7(a). The velocity profile is solved by combining equations (7) and (10)

$$0 = \rho_E \nabla \Phi + \mu \nabla^2 \mathbf{u}_{\mathbf{EOF}}, \tag{18}$$

where the electric charge density is obtained from equation (11). Considering the fluid behavior, the electric potential Φ is divided into two parts in Cartesian coordinates

$$\Phi = \phi(x) + \psi(y, z), \tag{19}$$

where ϕ is the electric potential due to the external electric field, $E = \partial\phi/\partial x$. Assuming that the electric body force is induced by a uniform zeta potential, ζ, equation (18) is integrated subject to the boundary conditions $\partial u/\partial y = \partial u/\partial z = 0$ at $y = z = 0$, and $u = 0$ and $\psi = \zeta$ at $y = \pm w/2$ and $z = \pm h/2$. The velocity profile is thus

$$u_{\mathrm{EOF}} = -\frac{\varepsilon\zeta E}{\mu}\left(1 - \frac{\psi}{\zeta}\right). \tag{20}$$

This equation indicates that the velocity is uniform in the region where the potential with the electric double layer is zero. Hence, for the thin electric double layer regime $\kappa L_c >> 1$, equation (20) reduces to the well-known Helmholtz–Smoluchowski equation,

$$u_{\mathrm{EOF}} = -\frac{\varepsilon\zeta E}{\mu}. \tag{21}$$

Because of the uniform velocity profile, a lot of microfluidic systems utilize electroosmotic flows, since they are free from the sample dispersion. However, in nanofluidic systems, the systems are often in the electric double layer overlap regime, $\kappa L_c < 1$.[13,14,24–26] This velocity profile is not uniform and the flow rate is significantly decreased. Furthermore, the chemical equilibrium and the ion adsorption on the wall must be considered to accurately predict the zeta potential.

Electrophoresis is the electromigration of molecules and particles suspended in the fluid, as illustrated in Figure 4.7(b). For dielectric particles of radius a, the electric flux lines are distorted by the presence of other particles, and the electrostatic force exerted on the surface charges balances with the fluid drag due to Stokes' law and ion migration in the electric double layer. The electric flux lines are also affected by the surface conductance, which is caused by an excess of ions in the electric double layer. The effect of the surface conductance on the electric field is described by the Dukhin number

$$Du = \frac{K_s}{aK}, \tag{22}$$

where K_s is the surface conductivity and K is the conductivity in the bulk.[27] The electrophoretic velocity of the particle is expressed as

$$u_{\mathrm{EP}} = \frac{2}{3}\frac{\varepsilon\zeta}{\mu} f(\kappa a, Du)E, \tag{23}$$

where f is a function dependent on κa and Du. For the thin electric double layer regime $\kappa a >> 1$, $f \rightarrow 1.5$ and equation (23) reduces to the Helmholtz–Smoluchowski equation. On the other hand, it is difficult to estimate the electrophoretic motion for the thick electric double layer regime, $\kappa a < 10$, since the phenomenon is far more complicated.

Electrokinetically-driven flows are easily controlled by switching the electric field through the electrode array, and are useful for the miniaturization of systems with recent micro- and nanofabrication technologies. It has been used for various nanofluidic systems.[28–31] Attolitter-scale rapid fluid dispersing with time resolution of 10–100 ms could be realized by switching the electric field in 2-D extended nanochannels.[3] The most significant difference from microfluidic systems is the effect of the electric double layer overlap. The nonuniform electroosmotic velocities and the interaction with the surface yield new characteristics for the material transported from the microfluidic systems. Despite the complicated flow field, these characteristics have potential for numerous applications, such as rapid and multifunctional separation, and fluidic-based electric devices.

4.5. Conclusion and Outlook

In this chapter, fundamental and recent advances in fluidic control methods based on pressure, visous force, and electrokinetics for extended nanospace were introduced. Attolitter to picolitter fluidic control has been achieved in each method. Pressure-driven and shear-driven flow systems, without an electric field, can provide the general chemical operation in the integrated chemical devices. Although the flow fields are complicated due to the electrophoresis, electrokinetically-driven flows are still available for specific applications, such as separation and electric operation.

However, the basic nanofluidics have still not been explored. Specific liquid properties, such as viscosity increase, enhanced proton exchange, slipping, and the electrokinetic effect, may influence the flow velocity profile in extended nanospace for pressure-driven, shear-driven, and electrokienticall-driven flows. Futher, it is possible that the molecules and particles will not follow the fluid flow, even for pressure-driven flows, owing to the large electrokinetic effect or entropic effect. Therefore, basic research to reveal the fluid dynamics in extended nanospace is required to further develop fluidic control methods.

References

1. Mawatari K., Tsukahara T., Sugii Y., and Kitamori T. (2010), Extended-nano fluidic systems for analytical and chemical technologies, *Nanoscale*, **2**, 1588–1595.
2. Tsukahara T., Mawatari K., and Kitamori T. (2010), Integrated extended-nano chemical systems on a chip, *Chem Soc Rev*, **39**, 1000–1013.
3. Kovarik M.L. and Jacobsen S.C. (2007), Attoliter-scale dispensing in nanofluidic channels, *Anal Chem*, **79**, 1655–1660.

4. White F.M. (1986), *Fluid Mechanics*, 2nd Ed., New York: McGraw-Hill.
5. Deen W.M. (1998), *Analysis of Transport Phenomena*, New York: Oxford University Press.
6. Neto C., Evans D.R., Bonaccurso E., Butt H.-J., and Craig V.S.J. (2005), Boundary slip in Newtonian liquids: a review of experimental studies, *Rep Prog Phys*, **68**, 2859–2897.
7. Voronov R.S., Papavassiliou D.V., and Lee L.L. (2008), Review of fluid slip over superhydrophobic surfaces and its dependence on the contact angle, *Ind Eng Chem Res*, **47**, 2455–2477.
8. Cao B.-Y., Sun J., Chen M., and Guo Z.-Y. (2009), Molecular momentum transport at fluid-solid interfaces in MEMS/NEMS: a review, *Int J Mol Sci*, **10**, 4638–4706.
9. Papautsky I., Brazzle J., Ameel T., and Frazier A.B. (1999), Laminar flow behavior in microchannels using micropolar fluid theory, *Sensor Actuator*, **73**, 101–108.
10. Hunter R.J. (1981), *Zeta Potential in Colloid Science*, London: Academic Press.
11. Bhattacharyya S., Zheng Z., and Conlisk A.T. (2005), Electroosmotic flow in two-dimensional charged micro- and nanochannels, *J Fluid Mech*, **540**, 247–267.
12. Baldessari F. and Santiago J.G. (2008), Electrokinetics in nanochannels: part I. Electric double layer overlap and channel-to-wall equilibrium, *J Colloid Interf Sci*, **325**, 526–538.
13. Baldessari F. and Santiago J.G. (2008), Electrokinetics in nanochannels: part II. Mobility dependence on ion density and ionic current measurements, *J Colloid Interf Sci*, **325**, 539–546.
14. Tsukahara T., Kuwahata T., Hibara A., Kim H.B., Mawatari K., and Kitamori T. (2009), Electrochemical studies on liquid properties in extended nanospaces using mercury microelectrodes, *Electrophoresis*, **30**, 3212–3218.
15. Brody J.P., Yager P., Goldstein R.E., and Austin R.H. (1996), Biotechnology at low Reynolds numbers, *Biophys J*, **71**, 3430–3441.
16. Ichikawa N., Hosokawa K., and Maeda R. (2004), Interface motion of capillary-driven flow in rectangular microchannel, *J Colloid Interf Sci*, **280**, 155–164.

17. Ren C.L. and Li D. (2004), Electroviscous effects on pressure-driven flow of dilute electrolyte solutions in small microchannels, *J Colloid Interf Sci*, **274**, 319–330.
18. Wang M., Chang C.-C., and Yang R.-J. (2010), Electroviscous effects in nanofluidic channels, *J Chem Phys*, **132**, 024701-1–024701-6.
19. Tamaki E., Hibara A., Kim H.-B., Tokeshi M., and Kitamori T. (2006), Pressure-driven flow control system for nanofluidic chemical process, *J Chromatogr A*, **1137**, 256–262.
20. Tsukahara T., Mawatari K., Hibara A., and Kitamori T. (2008), Development of a pressure-driven nanofluidic control system and its application to an enzymatic reaction, *Anal Bioanal Chem*, **391**, 2745–2752.
21. Desmet G. and Baron G.V. (2000), The possibility of generating high-speed shear-driven flows and their potential application in liquid chromatography, *Anal Chem*, **72**, 2160–2165.
22. Clicq D., Pappaert K., Vankrunkelsven S., Vervoort N., Baron G.V., and Desmet G. (2004), Shear-driven flow approaches to LC and macromolecular separations, *Anal Chem*, **76**, 430A–438B.
23. Vankrunkelsven S., Clicq D., Cabooter D., Malsche W.D., Gardeniers J.G.E., and Desmet G. (2006), Ultra-rapid separation of an angiotensin mixture in nanochannels using shear-driven chromatograpy, *J Chromatogr A*, **1102**, 96–103.
24. Qiao R. and Aluru N.R. (2003), Ion concentrations and velocity profiles in nanochannel electroosmotic flows, *J Chem Phys*, **118**, 4692–4700.
25. Pennathur S. and Santiago J.G. (2005), Electrokinetic transport in nanochannels. 1. Theory, *Anal Chem*, 77, 6772–6781.
26. Pennathur S. and Santiago J.G. (2005), Electrokinetic transport in nanochannels. 2. Experiments, *Anal Chem*, 77, 6782–6789.
27. Lyklema J. (1995), *Fundamentals of Interface and Colloid Science: Volume II Solid-liquid Interfaces*, San Diego: Academic Press.
28. Tegenfeldt J.O., Prinz C., Cao H., Chou S., Reisner W.W., Riehn R., Wang Y.M., Cox E.C., Sturm J.C., Silberzan P., and Austin R.H. (2004), The dynamics of genomic-length DNA molecules in 100 nm channels, *Proc Natl Acad Sci USA*, **101**, 10979–10983.
29. Garcia A.L., Ista L.K., Petsev D.N., O'Brien M.J., Bisong P., Mammoli A.A., Brueck S.R.J., and López G.P. (2005), Electrokinetic

molecular separation in nanoscale fluidic channels, *Lab Chip*, **5**, 1271–1276.

30. Vlassiouk I., Smirnov S., and Siwy Z. (2008), Ionic selectivity of single nanochannels, *Nano Lett*, **8**, 1978–1985.
31. Abgrall P. and Nguyen N.T. (2008), Nanofluidic devices and their applications, *Anal Chem*, **80**, 2326–2341.

Chapter 5

FUNDAMENTAL TECHNOLOGY: DETECTION METHODS

5.1. Single Molecule Detection Methods

Integration of chemical processes in small spaces provides many advantages for these processes, such as shorter analysis time, decreased sample and reagent volume, easier operation, and so on. Conversely, it requires sophisticated detection technologies due to the extended nanospace channels' small volumes and relatively short optical path lengths. Even in microspace, detection is sometimes a challenging issue. For example, the time-averaged amount of analyte in 1 μm^3 (1 fL) of a 1 nM fluid is 10^{-24} mol (Figure 5.1), less than a single molecule. If the dimensions of the detection volume are decreased to 100 nm, single molecule detection is required for even 1 μM solutions, which is the typical concentration targeted in analytical chemistry. Therefore, the detection methods on microchips require single molecule sensitivity. So far, several sensitive detection methods have been developed with single molecule sensitivity, and these methods are mostly dominated by optical and electrochemical methods. Optical methods especially have several advantages.

5.1.1. *Optical detection methods*

Optical detection methods are well known to be a very sensitive technique. If substrates of the microchips are transparent, optical detection allows *in situ* and non-invasive detection on chips with a very small volume; this is also an advantage of the optical detection methods. Optical detection usually utilizes the optical excitation of analyte molecules from electronic ground states to electronic excited states, as

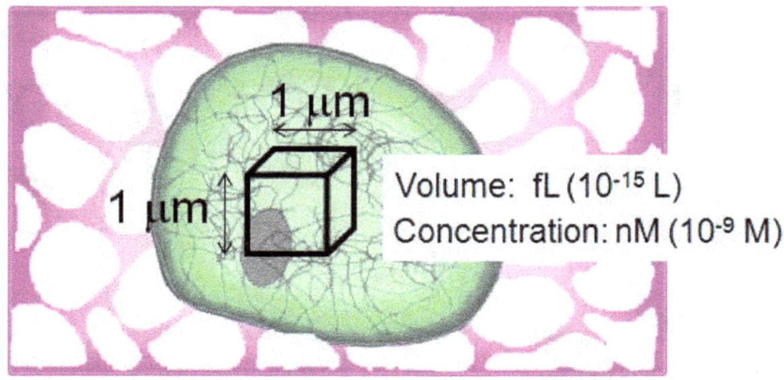

Figure 5.1. Requirement of single molecule sensitivity for detection methods in micro/extended nanospace.

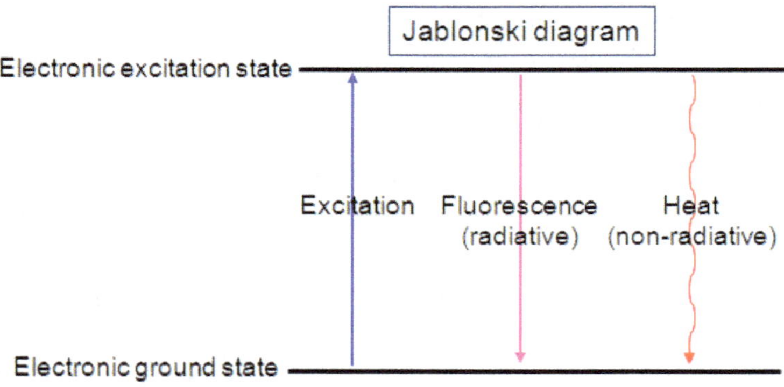

Figure 5.2. Illustration of Jablonski diagram.

shown in Figure 5.2, with a distribution in the vibrational states on each electronic state (not shown in Figure 5.2). The excited electron relaxes to the vibrational ground state on the electronic excited state in very short timescale (less than picoseconds). Then, two primary processes are observed for the relaxation: radiative relaxation and non-radiative relaxation.

When the non-radiative process occurs between singlet electronic states (i.e. no intersystem crossing), the radiated light is known as fluorescence. By coupling with a laser, which is called laser-induced

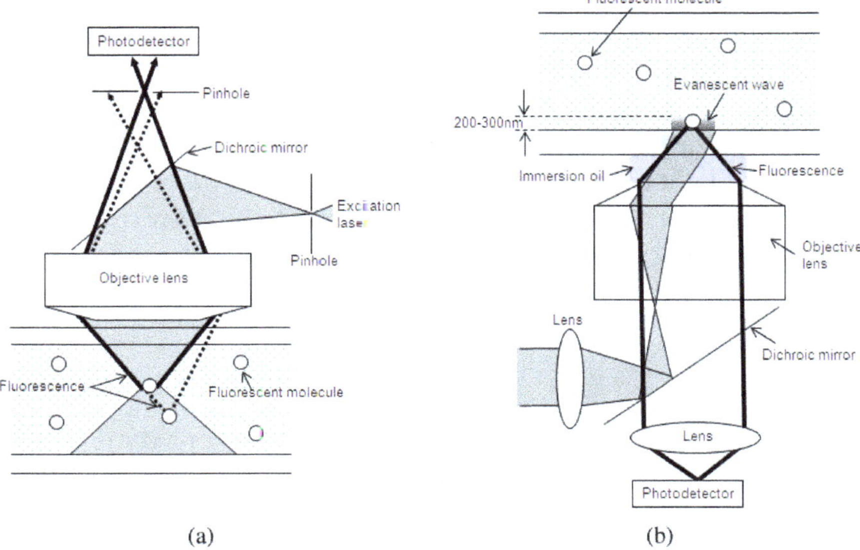

Figure 5.3. (a) Principle of confocal fluorescence microscopy and (b) Total internal reflection fluorescence microscopy.

fluorescence (LIF), very sensitive detection methods are realized, since laser beams allow high power density due to their coherence and the very narrow bandwidth of the wavelength. There are several optical detection schemes: typically confocal fluorescence microscopy and total internal reflection fluorescence microscopy are used, which have now been commercialized by several companies. In Figure 5.3, the basic optical scheme of confocal fluorescence microscopy and total internal reflection fluorescence microscopy are illustrated. In confocal fluorescence microscopy, an excitation laser is introduced to an objective lens through a dichroic mirror and focused onto the micro or extended nanospace channels. For a TEM_{00} Gaussian beam, the spot size, $2\omega_0$, of the excitation beam can be calculated by

$$w_0 = 0.6 \times \frac{\lambda}{NA}, \qquad (1)$$

where is the wavelength of the excitation beam is the numerical aperture of the objective lens. Fluorescent molecules inside the spot size are excited and release the energy as fluorescent light. The fluorescent light is collected by the same objective lens and focused to a pinhole. Finally, a sensitive photodetector detects the number of photons when a fluorescent burst occurs. If the fluorescent molecules are excited outside of the spot, only a very small portion of the generated fluorescent light can transmit through the pinhole and be detected by the photodetector. The depth resolution is almost restricted to the confocal length, z_c, which is calculated by

$$z_c = \frac{\pi \omega_0^2}{\lambda}. \tag{2}$$

As a result, fluorescent molecules in the confocal area can be detected. When an objective lens with high numerical aperture (NA) is used, the depth resolution reaches 200–300 nm. The lateral resolution is roughly estimated by the spot size. For accurate calculation, three-dimensional optical transfer functions should be applied. This detection configuration decreases background signals from solvents or impurities to realize a high signal-to-noise ratio, making single molecule detection possible. So far, many papers have reported on single molecule detection in microchannels by confocal fluorescent microscopy.[1]

While confocal fluorescent microscopy limits the detection volume by confocal detection scheme, total internal reflection restricts the excitation volume on the surface of the microfluidic channel. When the excitation beam is collimated by an objective lens and introduced to the bottom substrate of the microchip, with incident angle θ_i larger than θ_c (the critical angle for total internal reflection), the excitation beam is totally reflected on the liquid/substrate interface. An evanescent wave is simultaneously formed inside the liquid. The penetration depth, d_p, of the evanescent wave can be expressed as

$$d_p = \frac{\lambda}{4\pi\sqrt{n_1^2 \sin^2 \theta_i - n_2^2}}. \tag{3}$$

Where n_1 and n_2 are the refractive indices of the substrate and liquid, respectively. In the case of large incident angles, this approaches

100–200 nm. When fluorescent molecules enter the evanescent fields on the surface, fluorescence is generated and detected by the photodetector. Because of the very small volume of the evanescent wave, background signals from solvent and impurities can be greatly reduced, and single molecule detection becomes possible.[2]

LIF utilizes radiative decay processes from the electronic excited state. However, the radiative decay is a rather rare process, and almost all molecules release the absorbed energy as heat by non-radiative decay processes. Therefore, highly sensitive detection methods for heat generation are required for sensitive detection of a variety of nonfluorescent molecules. One such method is thermal lens microscopy (TLM). The basic principle is illustrated in Figure 5.4. In TLM, two laser beams are utilized: an excitation laser and a probe laser. The wavelength of the excitation beam is tuned to the light absorption band of the analyte molecules, while that of the probe beam is adjusted to avoid the absorption band. The excitation laser is tightly focused on the sample with size $2\omega_0$, which is calculated by equation (1). The analyte molecules in the sample absorb the excitation beam and relax via non-radiative relaxation processes. The relaxation time is typically in the ns to μs scale. Heat is then released

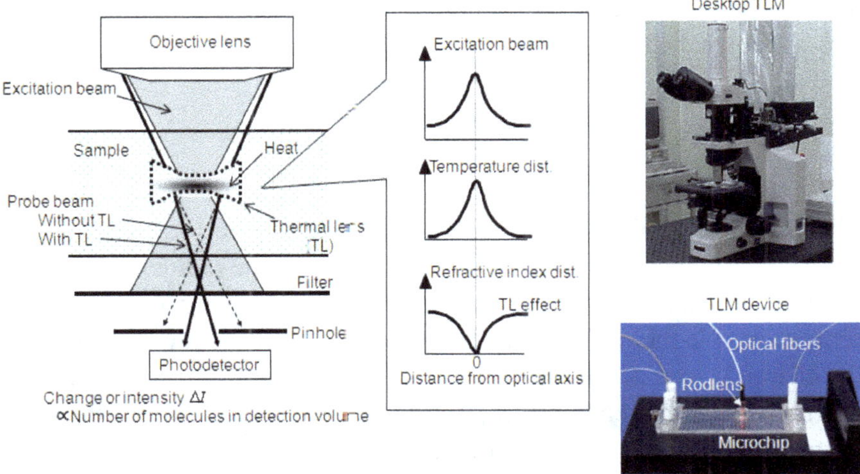

Figure 5.4. Basic principle of thermal lens microscopy (TLM) and its devices.

to the solvent around the molecule, and the temperature increases. This is called the photothermal phenomenon. The intensity distribution of the excitation beam has a Gaussian profile. The amount of light absorbed is linearly dependent on the intensity distribution, and the temperature distribution is similar to the intensity distribution. For continuous-wave lasers, the excitation beam is intensity-modulated by a chopper for lock-in amplifier detection. Because the thermal lens effect, illustrated below, is very small ($\Delta T < 10^{-3}$ °C), the extraction of the thermal lens signal by the lock-in amplifier is essential. In this case, the size of the temperature distribution is enlarged by thermal diffusion during each cycle. The thermal diffusion length μ_{th} is expressed as,

$$\mu_{\text{th}} = \sqrt{\frac{D}{\pi f}}, \tag{4}$$

where D is the thermal diffusivity of the medium, and f is the chopper frequency. In TLM, the chopper frequency f is around 1 kHz for achieving a high signal-to-noise ratio, and the thermal diffusion length μ_{th} becomes 7 μm, which is the characteristic length of the thermal lens. In paraxial theory (ray optics approximation), the thermal lens effect is treated as a simple concave lens.[3] The strength of the lens at thermal equilibrium is expressed as,

$$\frac{1}{F} = \frac{P\varepsilon Cl(dn/dT)}{\pi J\kappa\omega_0^2}, \tag{5}$$

where F is the focal length of the thermal lens, J is the Joule coefficient, κ is the thermal conductivity of the medium, ω is the radius of the beam waist, P is power, ε is the molar absorption coefficient, C is the concentration, l is the optical path length of the cell (usually the depth of the channel), and dn/dT is the first derivative of the refractive index, n, of the medium by temperature, T. As shown in this equation, focusing of the excitation beam makes the detection volume small (which means there is only a small background signal), and enhances the decree of thermal lens leading to a high signal-to-noise ratio for single molecule detection. Organic solvents usually show higher sensitivity than aqueous solutions due to the higher value of $(dn/dT)/\kappa$.

Next, the probe beam is coaxially introduced into the objective lens and focused on the sample. Without the excitation beam focused on the sample, the thermal lens effect cannot be induced, and the path of the probe beam is not changed. In contrast, when the excitation beam is focused on the sample, the thermal lens effect is induced, changing the optical path of the probe beam. The change in the optical path is detected by measuring the probe beam intensity through a pinhole. The intensity change ΔI is defined as the thermal lens signal. For dilute solutions, the equation

$$\Delta I \propto \frac{1}{F} = \frac{P\varepsilon Cl(dn/dT)}{\pi J\kappa\omega_0^2}. \tag{6}$$

Here, there is an important optical configuration to realize sensitive TLM, which is shown in Figure 5.5. When the excitation and probe beams are focused on the same position (no focus difference), the thermal lens does not affect the propagation of the probe beams, and no thermal lens signal is obtained. If a certain focus difference, Δz, is selected, depending on the optical conditions, very sensitive detection

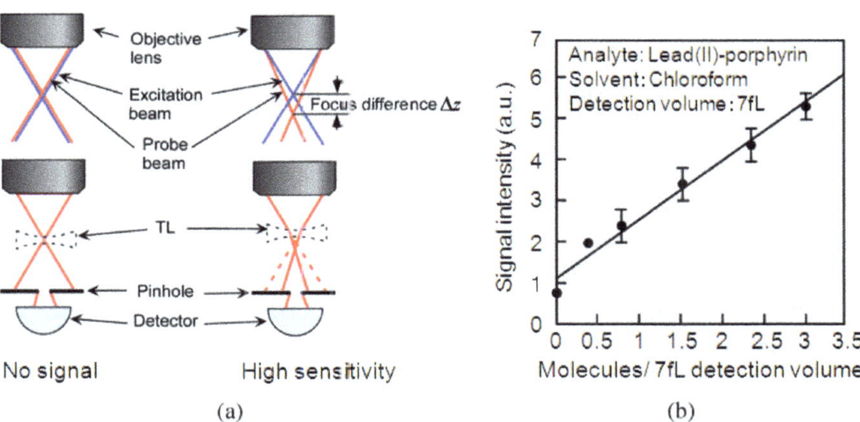

Figure 5.5. (a) Importance of focus difference between excitation and probe beams for realizing sensitive TLM and (b) Concentration determination at single molecule level.

becomes possible. The optimum value of Δz can be roughly estimated by

$$\Delta z = \sqrt{3} z_c . \tag{7}$$

In optimized conditions, concentration determination at single molecule level has been demonstrated.[4]

Recently, TLM was further improved for concentration determination in extended nanospace channels. Differential interference contrast optics were coupled with TLM optics (DIC-TLM), where the probe beam is separated to two beam spots by a DIC prism and the excitation beam is focused to one spot of the probe beam by adjusting the polarization direction of the excitation beam. When the analyte molecules absorb the excitation beam and generate heat, the refractive index of the solution decreases and the phase of the probe beam, coaxial with the excitation beam, changes. The phase difference of the two spots on the probe beam is detected by differential interference. This new detection scheme allows optically background-free detection, similar to LIF, and sensitive detection is realized. As a result, concentration determination of non-fluorescent molecules was realized for the first time at several-hundred molecules for aqueous solution. Further optimization can be expected to improve the sensitivity.[5,6]

Although detection becomes difficult when decreasing the analytical space, extended nanospace can provide unique properties for detection methods. For single molecule detection on a microchip, the laser beams are focused to the microchannel with a typical spot size of approximately 1 μm in diameter. The detection volume is in the fL (1 μm^3) scale, which is much smaller than the size of the microchannel (10–100 μm). The detection efficiency, i.e. the number of detected molecules divided by the number of totally introduced molecules, is thus very low (< 0.1%), however, the size of the extended nanospace channel is less than 1 μm, so detection of all the analyte molecules can be expected, as illustrated in Figure 5.6. These features are actually proven for LIF (single molecule detection) and TLM (single nanoparticle detection).[7,8]

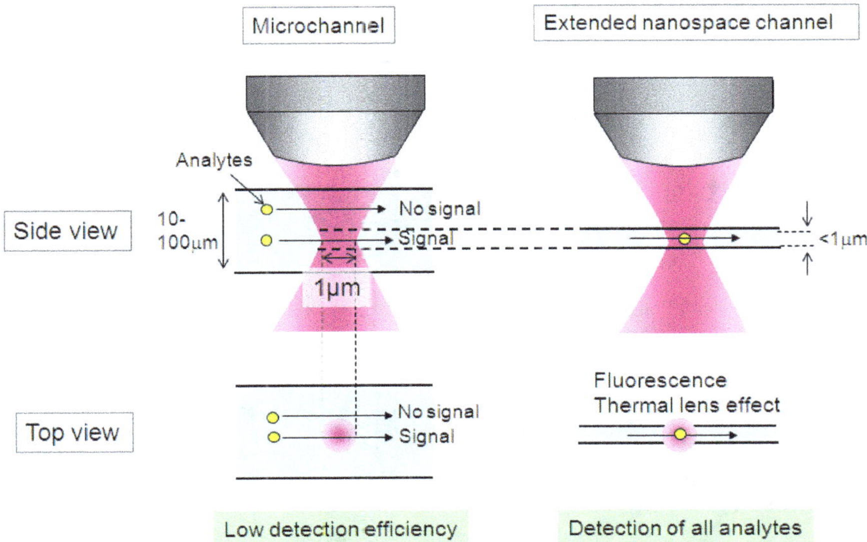

Figure 5.6. Detection of all analytes by extended nanospace channels.

5.1.2. *Electrochemical methods*

Electrochemical methods are another candidate for single molecule detection. The first single molecule detection was realized by Bard and coworkers.[9] The group developed a scanning electrochemical microscope (SECM), which measured current through electrochemical reactions. The gaps between the electrodes were made quite small (~10 nm), and the detection volume was reduced to 10^{-21} L. Due to the very small space, the diffusion time of the analyte molecules was reduced to ~100 ns scale, allowing high speed and repeated reduction-oxidation reactions by the diffusion alone. As a result, a detectable current (~pA) was obtained from the reduction-oxidation reactions of the single molecules. However, the method was limited to molecules with redox electrochemical reaction activity.

Recently, nanopore is increasingly utilized for sensitive detection. The principle is illustrated in Figure 5.7. A membrane with small pores (of diameters of d_b and d_t) is mounted between two electrodes. A certain voltage is applied across the electrodes, and a current is

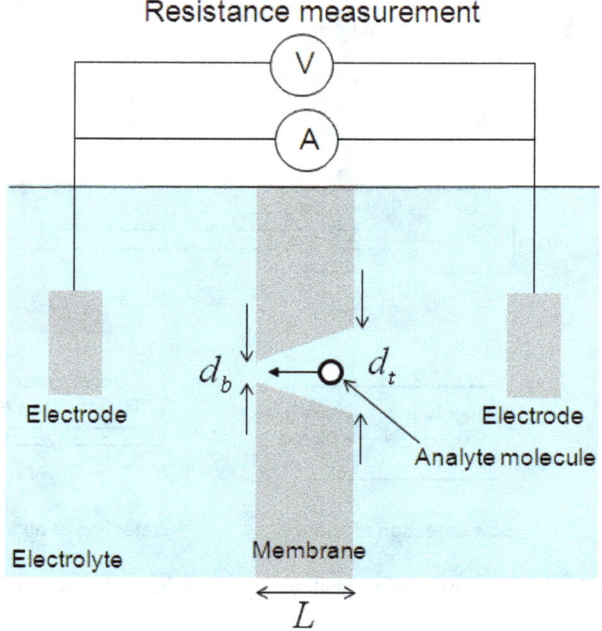

Figure 5.7. Principle of single molecule detection by nanopore-based resistance measurement.

induced through the nanopore. In particular, d_b is as small as several nanometers to achieve single molecule detection. In this case, the resistance can be expressed by

$$R = \frac{4\rho L}{\pi d_b d_t}, \tag{8}$$

where ρ is the specific resistance of an electrolyte. From equation (8), a decrease in the pore size leads to high resistance, which in turn leads to a large current change when single molecules pass through the nanopore. Actually, by measuring the resistance from the current–voltage curve, a single porphyrine molecule can be successfully detected.[10] The principle is very simple, and its applicability is wide. Although the coupling with micro- or extended nanospace channels is still an issue, due to the difficulty of integration, some groups recently reported the fabrication of nanopores by top-down technologies.[11]

5.2. Measurement of Fluidic Properties

Understanding flow structures confined in extended nanospace is important to develop novel, miniaturized chemical and biological systems. Previous work has reported different flow properties of this scale from bulk fluid. Kitamori *et al.* reported a capillary flow velocity in 250 nm channel, which indicated higher water viscosity than the bulk.[12] Some studies have dealt with the electroviscosity in small channels generated by a near-wall ion layer (electric double layer).[13,14] Slip velocity due to hydrophobic wall surface is also a factor affecting the flow structure in small channels.[15,16] Therefore, experimental approaches are required to study fluid dynamics and material transport, however, measurement of flows in such a small scale is difficult due to its low flow rate and small size. This section provides fundamental and recent advances in flow measurement techniques, which can be used for extended nanospace channel flows. Both nonintrusive techniques and tracer-based techniques are described. The nonintrusive methods based on electrokinetics have measurement resolutions of 10^{-11} L min^{-1}. The tracer-based techniques, using evanescent waves and super resolution microscopy, can achieve nanometer-order spatial resolution in the flow imaging.

5.2.1. *Nonintrusive flow measurement techniques*

Nonintrusive flow measurement techniques, without flow tracers like fluorescent molecules and particles, are essential to investigate flows, especially small scale flows of large surface-to-volume ratio. They have no influence on the fluid and surface properties. Although they cannot be applied to flow visualization, they have more flexibility and versatility than tracer-based measurement techniques. In the case of macroscale flows, with a length scale of order 1 m, a typical nonintrusive technique is using mechanical-based flow meters with a resolution in the order of Lh^{-1}. For microscale flows, several researchers have evaluated the flow rate in microchannels by measuring the mass of water at the channel output.[13,17,18] This method is simple in principle and provides reliable results, with an accuracy totally dependent on the resolution of a balance of approximately 10^{-5}–10^{-3} g. On the other hand, in extended nanoscale regimes, the flow rate is around 10^{-11}–10^{-7} L min^{-1} in volume, and 10^{-14}–10^{-10} g min^{-1} in weight of water, on an assumption that the average velocity is 1 mms^{-1}. Therefore, it is quite difficult to obtain the flow rate in the extended nanospace channel using these techniques, even when the experiment is conducted with hundreds of channels and the flow rate is magnified to 10^{-9}–10^{-5} L min^{-1} (10^{-12}–10^{-8} g min^{-1}). In order to measure nanochannel flows, several studies have used methods based on electrokinetics.

5.2.1.1. *Streaming potential/current measurement in pressure-driven flows*

A measurement technique using streaming potential/current in electrolyte solutions, which is generated by pressure-driven flow, has been applied for nanochannels.[19–21] The streaming potential/current measurement method is a traditional technique to evaluate surface properties such as the zeta potential.[22–24] The theory for the generation of electric signals, with regard to the surface conditions and fluid flow, is well established. Figure 5.8 illustrates a schematic of the

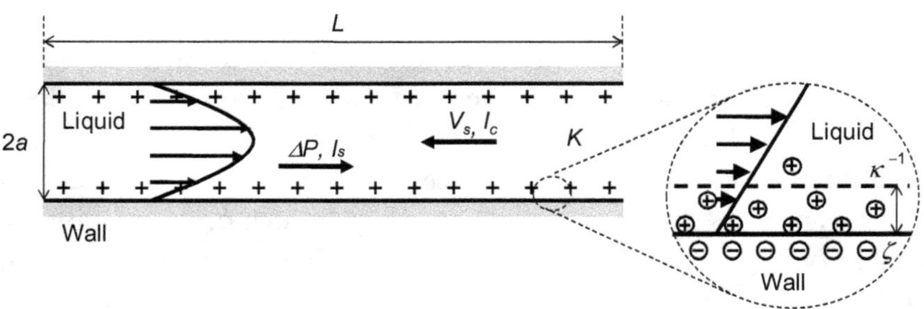

Figure 5.8. Schematic illustration of streaming potential/current generated by pressure-driven flow.

streaming potential/current. Generally the surface charge and accumulated counter-ions in the proximity of the surface form an electric double layer of nanometer-order thickness. An application of pressure induces the fluid flow and leads to an ion flow in the electric double layer. This pressure-driven ion flow in turn generates a streaming current, I_s, that can be calculated as[25]

$$I_s = -\frac{\pi a^2 \varepsilon \zeta}{\mu L}\left[1 - \frac{2}{j\kappa a}\frac{J_1(j\kappa a)}{J_0(j\kappa a)}\right]\Delta p, \tag{9}$$

where a is the channel radius, ε is the dielectric constant, ζ is the zeta potential (surface electrostatic potential), μ is the viscosity, L is the channel length, κ^{-1} is the Debye length (thickness of the electric double layer), J_0 and J_1 are Bessel functions of zero and first order, respectively, and Δp is the pressure difference across the channel. The ion flow yields the accumulation of charge and induces an electric field. This field causes a current flow in the opposite direction to the pressure-driven flow. A steady state is achieved when this conduction current, I_c, is equal to the streaming current. The resulting electric potential difference across the channel is the streaming potential

$$V_s = \frac{\varepsilon \zeta}{\mu K}\left[1 - \frac{2}{j\kappa a}\frac{J_1(j\kappa a)}{J_0(j\kappa a)}\right]\Delta p, \tag{10}$$

where K is the electrical conductivity.

Recently, Decker *et al.* established the method in 1-D extended nanospace,[20] while Wang *et al.* reported measurements in nanopores.[21] However, the investigation of specific liquid properties in 2-D extended nanospace is difficult due to the difficulty of size-controlled fabrication and accurate pressure-driven fluidic control. Kitamori *et al.* developed a method for measuring streaming potential/current in size-controlled 2-D extended nanospace on a glass substrate.[19] An experimental setup with precise flow control and sensitive detection was developed, as shown in Figure 5.9. A fused-silica chip

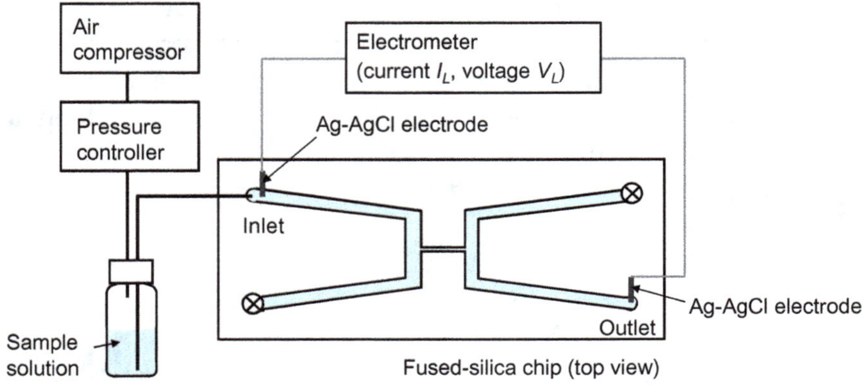

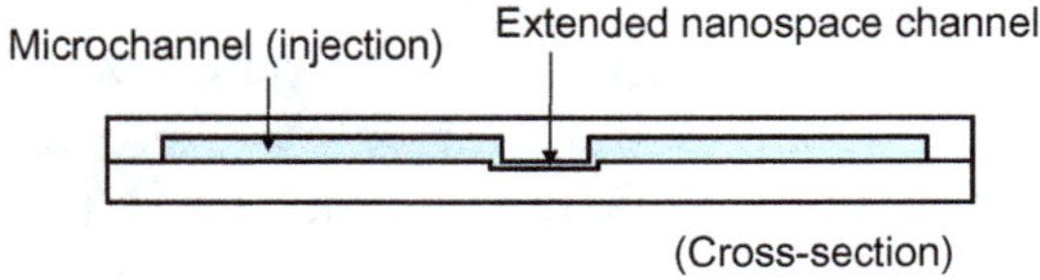

Figure 5.9. Schematic illustration of the streaming potential/current measurement system.

of extended nanospace channels, combined with microchannels for injection, was prepared for the measurements. The extended nanospace channels were fabricated on a fused-silica plate by EB lithography and plasma etching. The microchannels were etched on a separate glass slide, and the glass plates were then combined by thermal fusion bonding. The pressure drop and the electrical resistance of microchannels are negligible compared to those of nanochannels. The sample solution was stably driven with an air-pressure controller of order 0.1 MPa. The current generated from the inlet to the outlet, $I_L = I_s$, and the potential at the outlet relative to the inlet, $V_L = -V_s$, were measured by an electrometer through Ag-AgCl electrodes. A linear relationship between the current/voltage values and applied pressures was obtained at the steady state, as shown in Figure 5.10. Considering the Hagen–Poiseuille relation (see Section 4.2 equation (15)), these applied pressures correspond to a flow rate of order 10^{-11} L min^{-1}. Therefore, this system is a useful tool for sensitive

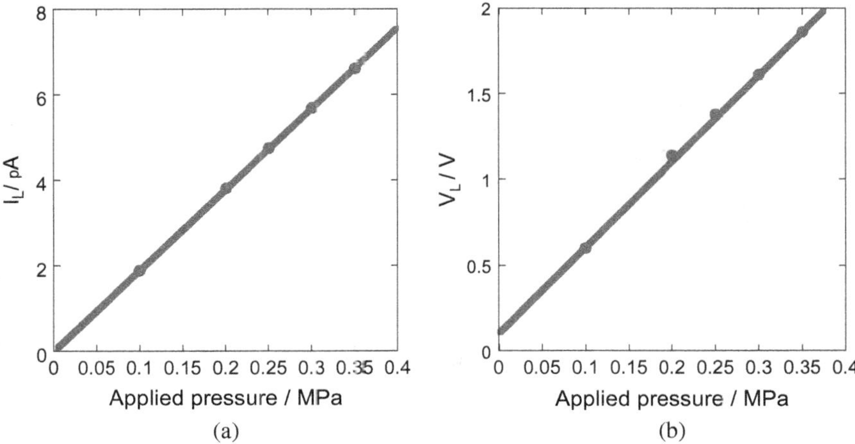

Figure 5.10. (a) Current values and (b) Voltage values as functions of applied pressure in an extended nanospace channel of 610 nm × 560 nm. Sample is pure water (Redrawn from Ref. 19).

detection of surface chemistry and liquid properties in extended nanoscale, and also for the flow rate measurement if a calibration between the current/voltage signals and the flow rate is successfully conducted.

5.2.1.2. *Current monitoring in electroosmotic flow*

The current monitoring technique is a method used to measure the electroosmotic flow rate.[26] This method has been used to investigate both the electroosmotic mobility and the surface zeta potential, with different materials and liquid properties in microchannels.[27–29] Figure 5.11 shows a schematic of the current monitoring technique. An electrolyte solution of low concentration is prepared by diluting the original sample solution, and injected into a channel from an outlet reservoir. Conversely, an inlet reservoir is filled with the original solution (high concentration). The pressure difference between the inlet and outlet reservoirs is adjusted to zero. Electroosmotic flow is induced by an electric field, and the solution in the channel is exchanged from the low concentration to the high concentration

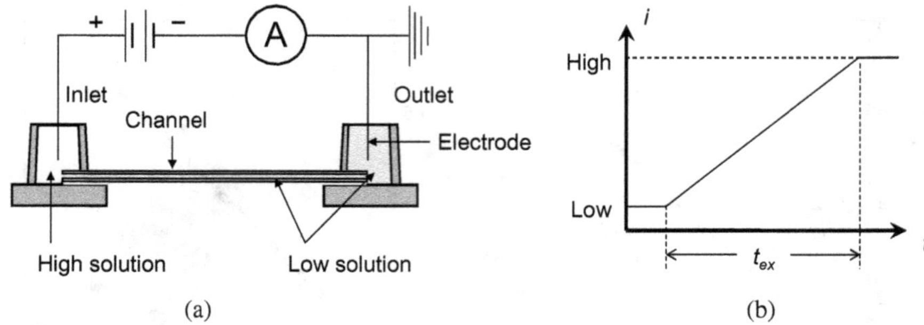

Figure 5.11. (a) Schematic illustrations of the current monitoring, when a liquid of low concentration is exchanged for that of high concentration, (b) Electric current as a function of time when exchanging the solution.

solution. Since the electrical conductivity is proportional to the electrolyte concentration, the current in the channel is increased with the solution exchange, as depicted in Figure 5.11(b). Hence the velocity of electroosmotic flow, u_{EOF}, is calculated from

$$u_{\mathrm{EOF}} = \frac{L}{t_{\mathrm{ex}}}, \tag{11}$$

where t_{ex} is the time for exchanging the solution.

Several studies have investigated nanochannel flows using the current monitoring technique. van der Berg and coworkers developed the current monitoring system for silicon nitride 1-D nanochannels with exchanging 100 mM and 50 mM buffers.[30] Their system successfully achieved electric current monitoring of order 10^{-9} A and an electroosmotic flow rate of 10^{-12} L min^{-1}. Pennathur and Santiago conducted measurements for fused-silica nanochannels of 2052 nm, 102 nm, and 38.5 nm depths.[31] The results of these measurements are in good agreement with an analytical model presented by the authors.[32] The electroosmotic velocity decreases strongly with the increasing ratio of the Debye length to the channel dimension $1/\kappa a$. This technique is available for measuring electroosmotic flow in nanochannels and investigating nanoscale electrokinetics.

5.2.2. *Optical flow imaging techniques using a tracer*

Optical measurement techniques using a tracer have been widely used for small scale flows. The flow tracer is seeded into the fluid, and the velocity is obtained from the tracer displacement in the flow. Here, fundamentals and recent advances of measurement techniques using fluorescent dye or particles are provided.

5.2.2.1. *Properties of flow tracers*

For the flow visualization, tracers such as molecules and particles are seeded into the fluid, and the flow velocity is obtained based on an assumption that the tracer follows the flow. The size of the tracer is often in the order of nanometers, which is smaller than the light wavelength in the order of 100 nm, to achieve sufficient spatial resolution in micro- and nanoscale flow measurements. Since the scattering light in this regime is weak for the detection, tracers, which can emit fluorescence, are broadly used for the measurements.

One of the problems in small scale flow visualization is that the tracer suspended in the fluid is subject to significant diffusion, described by the Stokes–Einstein relation:

$$D = \frac{k_b T}{3\pi\mu d}, \tag{12}$$

where D is the diffusion coefficient, k_b is the Boltzmann constant, T is the temperature of the fluid, μ is the fluid viscosity, and d is the particle diameter. Hence the measured velocity is a superposition of the fluid velocity and the diffusive velocity, with mean square displacement $\langle x^2 \rangle$

$$\langle x^2 \rangle = 2D\Delta t, \tag{13}$$

where Δt is the time resolution of the measurement. In order to conduct the measurement for extended nanoscale, the tracer should be at least less than 10^{-6} m, however, the diffusion displacement becomes larger with decreasing tracer size. In particular, when the tracer size is less than 10 nm, the diffusion scale is often greater than the channel size. In this case, the spatial resolution of the measurement is defined by $\langle x^2 \rangle^{1/2}$ rather than d since the tracer has information from the region it passes through during Δt. A short time interval can reduce $\langle x^2 \rangle^{1/2}$, but a measurement of Δt is also required to maintain accuracy. Therefore, the tracer diameter and the diffusion displacement is a tradeoff to obtain sufficient spatial resolution and accuracy in the measurement.

Another problem for flow visualization in extended nanoscale may be tracer adsorption on the wall. Since the surface-to-volume ratio is large, and dominant in this scale, tracer adsorption can significantly affect the flow. Therefore, properties of the tracer, such as electric charge and chemical affinity, should be optimized to avoid adsorption on the wall, with a lower tracer concentration also desirable.

5.2.2.2. *Scalar image velocimetry*

Scalar image velocimetry (SIV) is a measurement technique using a transported scalar such as fluorescent dye. A scalar band is formed and injected into the channel as the flow tracer and then transported through the channel by advection and diffusion. The velocity is obtained from the displacement of the band in a sequence of images, as shown in Figure 5.12. Furthermore, this technique is able to measure the diffusion coefficient of the scalar. When the band is injected into quiescent fluid, the scalar is transported only by the diffusion. A concentration profile of the scalar as a function of time is given simply by the Gaussian distribution

$$c = \frac{N_0}{\pi a^2} \frac{1}{2\sqrt{\pi D t}} \exp\left(-\frac{x^2}{4Dt}\right), \tag{14}$$

where N_0 is the number of moles of scalar injected. Therefore, the diffusion coefficient is estimated after obtaining the Gaussian fit of the scalar image based on the assumption that the image intensity is

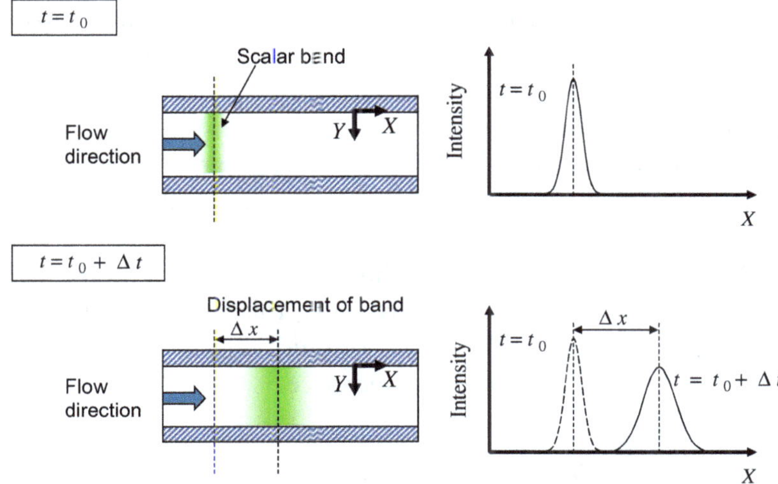

Figure 5.12. Schematic illustrations of the scalar image velocimetry.

proportional to the concentration. Experimentally, the scalar band is formed chemically (e.g. caged fluorescent dye,[33] phosphorescence,[34] photobleaching,[35] etc.) or mechanically (switching of the flows in a cross-section channel). One of the advantages of this technique is that the tracer consists of molecules with sizes of order 10^{-9} m, which are suitable for extended nanospace.

For electroosmotic flows, Pennathur and Santiago conducted measurements of 1-D nanochannel flows by epi-fluorescence imaging. Rhodamine B, bodipy, and fluorescein were used as fluorescent tracers with ion valences of 0, –1 and –2, respectively.[31] The scalar band was injected by an external electric field. The results show that the average velocity of the electrically neutral dye agrees with a fluid velocity obtained by current monitoring. On the other hand, the charged dyes have an electrophoretic migration component in addition to advection. In nanochannels with electric double layer overlap, especially, nonuniform distribution of charged dyes by electrostatic potential also affects the average velocity, owing to a nonuniform electroosmotic flow profile.

Desmet *et al.* investigated the diffusion coefficients of the molecules in 1-D nanochannels. A scalar band of fluorescent dyes and fluorescently-labeled ssDNA molecules were injected into the nanochannel by a shear-driven flow system, which was developed by the authors' group. Then measurements were conducted in a quiescent fluid. The results revealed that the diffusion coefficients of molecules decrease in nanochannels compared to the bulk values, and this reduction is relatively stronger for larger molecules.[36]

SIV cannot be applied to obtain a velocity profile of nanochannel flows due to the optical diffraction limit and significant molecular diffusion, as mentioned above. Nevertheless, this technique is available to investigate the apparent behavior of fluids and molecules interacting with the surface in extended nanospace.

5.2.2.3. *Nanoparticle image velocimetry*

Particle image velocimetry (PIV) is a two-dimensional velocity measurement technique using small tracer particles. The tracer particles are seeded into the fluid and images of the test section are captured by a digital imaging device. The fluid velocity is obtained from the displacement of the tracer particles during a short time interval. Microparticle image velocimetry (μPIV), i.e. PIV combined with microscopy, has been broadly used for microscale flow visualization.[37,38] However, the maximum spatial resolution is limited to the optical diffraction limit, which is comparable to the light wavelength. In addition, most μPIV studies use submicron tracer particles. Therefore, it is difficult to apply this technique to the extended nanoscale.

In order to achieve spatial resolution over the diffraction limit, a particle-based velocimetry using the evanescent wave with total internal reflection of light, i.e. nanoparticle image velocimetry (nPIV), has been proposed to achieve a nanometer-order spatial resolution of flow visualization.[39,41] In nPIV, the evanescent wave is used to illuminate fluorescent nanoparticles. This technique can measure the velocity within the first several hundreds of namometers next to the wall. Figure 5.13(a) shows a schematic of the near-wall velocimetry by the evanescent wave illumination. The fluorescent tracer particles within a distance of order 10^{-7} m of the wall are illuminated by the evanescent wave, and

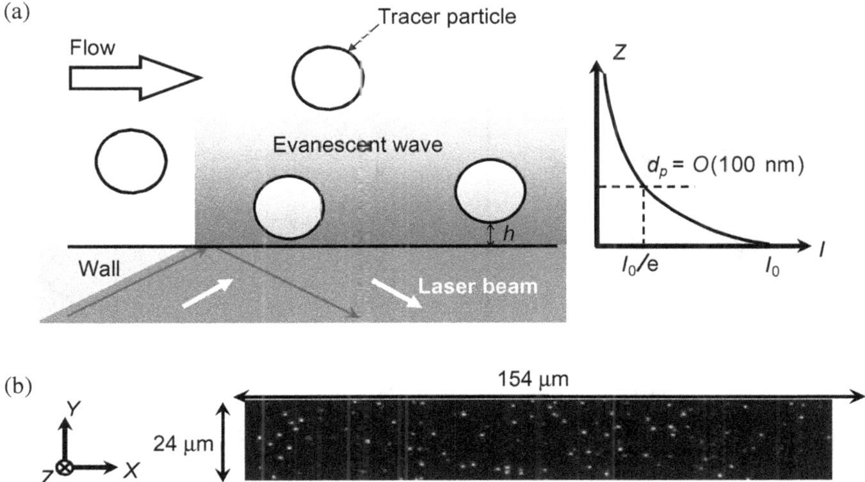

Figure 5.13. (a) Schematic illustration of the near-wall velocimetry and (b) a fluorescence image of 100 nm particles illuminated by an evanescent wave, captured by an EMCCD camera through a 63× objective lens of a numerical aperture of 0.7 and 0.5× lens (Adopted from Ref. 42).

images of the *XY*-plane are acquired by an imaging device, as shown in Figure 5.13(b).

PIV measures the fluid velocity u by image processing to determine the particle displacement Δx during the frame interval Δt. The processing algorithm can be divided into two types: particle image velocimetry (PIV) and particle tracking velocimetry (PTV), as shown in Figure 5.14. In PIV, a pattern of tracer particles in the first frame ($t = t_0$) is extracted as an interrogation window and used as a reference for the match candidates in the second frame ($t = t_0 + \Delta t$). The most likely particle–image pairing between the first and second frame is determined. On the other hand, the method to extract the velocity using the displacement of individual tracer particles is termed PTV. The location of particles in each image is detected, and the displacements of the individual particles are then obtained by matching the particles in the first image to those in the second image.

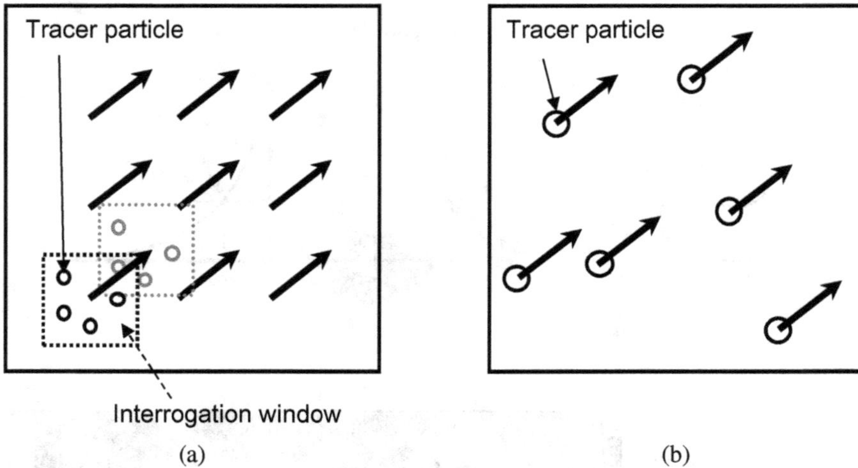

Figure 5.14. Schematic illustrations of (a) particle image velocimetry and (b) particle tracking velocimetry.

However, small tracer particles suspended in the fluid are subject to significant Brownian diffusion as described above. In order to reduce the error by the Brownian motion, ensemble averaging of the velocities, e.g. time averaging, cross-correlation averaging, and local spatial averaging, is applied.[42,43] Since the Brownian diffusion is spatially and temporally random, the average displacement of the Brownian diffusion is identically equal to zero. Therefore the fluid velocity can finally be obtained.

The evanescent wave intensity has an exponential decay, which is a function of the distance from the wall; as such, one can determine the distance between the particle edge and the wall, h, using the fluorescent intensity of particles (Figure 5.13(a)). Li and Yoda measured the near-wall velocity profile of Poiseuille flow through a fused-silica microchannel with dimensions of 41 μm × 469 μm. Fluorescent polystyrene particles with a diameter of 100 nm, were used as a tracer.[41] The displacements of the particles during $\Delta t = 1.5$ ms were obtained by PTV, and the particle position at the distance from the wall, $z = h + d/2$, was determined using the particle image intensity. Figure 5.15(a) shows particle number density c as a function

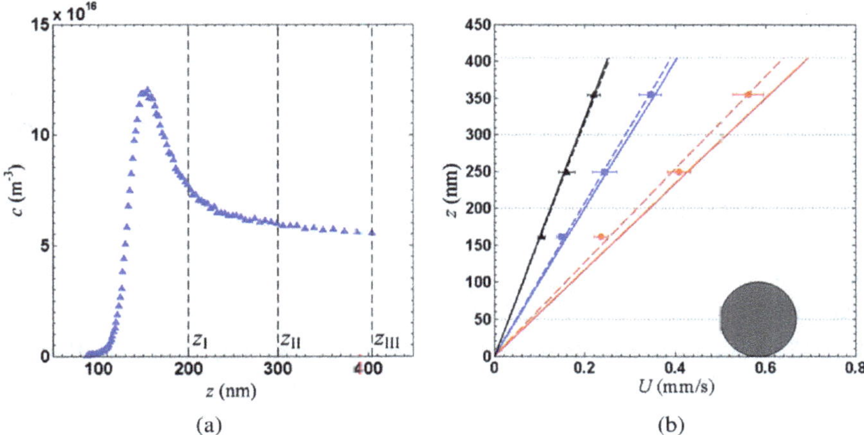

Figure 5.15. (a) Particle number density c as a function of z the wall-normal distance for the 10 mM sodium tetraborate solution ($Na_2B_4O_7$), (b) Velocity U as a function of z compared with the expected velocity profile (solid line) at the shear rate $G = 620\ s^{-1}$ (▲), $1000\ s^{-1}$ (■) and $1720\ s^{-1}$ (●). The dashed lines denote a linear-curve fit of the experimental data points and the origin. The error bars represent the 95% confidence level (Adopted from Ref. 41).

of z. The particles have a nonuniform distribution due to the electrostatic repulsive force between negatively charged surfaces and the attractive van der Waals force. The measurement region was then divided into three layers of different z-ranges, and the ensemble-averaged velocities for each layer were calculated. The location of the velocity data in each layer was corrected using the centroid of nonuniform particle distribution. Figure 5.15(b) shows the near-wall velocity profile as a function of z within 400 nm for different shear rates. The velocity profile of the Poiseuille flow agrees with the expected profile (a solid line). In addition to the velocity profile and the particle distribution, this technique can measure the near-wall diffusion. Huang and Breuer measured the near-wall hindered Brownian diffusion of 1.5 μm particles.[44] The results show that the diffusion coefficients are decreased near the wall in good agreement with the classical theory based upon the hydrodynamic wall effect.

A strong tool to investigate the fluid flow and colloid dynamics within a distance of order 10^{-7} m from the wall is nPIV. However, for investigating extended nanospace channel flows, it still has several problems. Although this technique achieves order 10^{-8} m resolution in the *Z*-direction, the resolution in the *XY*-plane is the same as μPIV. Furthermore, current work uses tracers of 100 nm diameter as a minimum, but this size is too large for the extended nanospace channel. Guasto and Breuer have developed an evanescent wave-based velocimetry using a quantum dot of hydrodynamic diameter 17 nm, but it is difficult to obtain the near-wall velocity profile, mainly because of the significant diffusion.[45] Therefore further development of the method is required.

5.2.2.4. *Laser-induced fluorescence photobleaching anemometer with stimulated emission depletion*

In contrast to the above techniques using tracer images, point velocity measurement techniques (e.g. laser Doppler velocimetry,[46] fluorescence correlation spectroscopy,[47] and fluorescence recovery after photobleaching[48]) have been developed for small flow measurements. Wang has developed a velocity measurement technique using fluorescence, i.e. laser-induced fluorescence photobleaching anemometer (LIFPA).[49] This technique is based upon the photobleaching of fluorescent dye flowing through a laser beam spot, as illustrated in Figure 5.16. The advantages of LIFPA compared to other point measurement methods are the simple optical system, high time resolution, and tracer of molecular size, which is suitable for nanochannel flows.

The photobleaching occurs when a high-intensity light source chemically changes a fluorophore to not participate in the absorption/emission process. The fluorescent intensity *I*, reduced by the photobleaching, is given by an exponential decay function

$$I = I_0 \exp\left(-\frac{t}{\tau_p}\right), \tag{15}$$

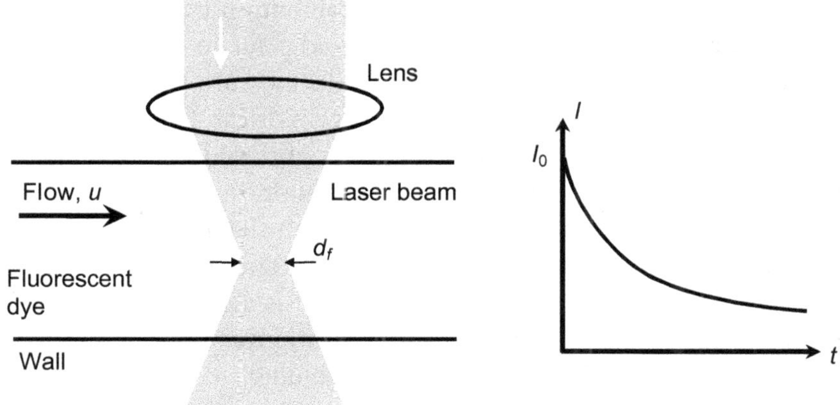

Figure 5.16. Schematic illustration of LIFPA. A fluorescent dye solution in a channel passes through a laser beam with photobleaching. The decaying time for photobleaching is the residence time of the dye in the laser beam.

where I_0 is the fluorescent intensity at $t = 0$ and τ_p is the time constant of photobleaching. When the fluorescent dye solution passes through the laser spot, of width d_f, in the flow field of velocity u, the average residence time of the dye in the laser spot is given by

$$t = \frac{d_f}{u}. \tag{16}$$

Then the fluorescent intensity is related to the fluid velocity as

$$I = I_0 \exp\left(-\frac{d_f}{u\tau_p}\right). \tag{17}$$

Therefore the velocity can be obtained by fluorescence detection. This equation indicates the fluorescent intensity increases with increasing velocity. In reality, the photobleaching time constant depends on various parameters such as laser intensity, optical system, and dye and liquid properties. Hence a calibration between the fluorescent intensity and the velocity is required prior to measurement.

Wang and coworkers applied LIFPA to measure the velocity distribution of pressure-driven flow in cylindrical and rectangular microchannels.[50] The velocity profile was obtained by scanning the detection spot with a 3-D piezoelectric nanopositioning stage. The detection volume of 200 nm and 400 nm in the lateral and axial direction of the laser beam, respectively, was achieved by confocal microscopy. However, this spatial resolution is still not sufficient for extended nanospace channels. Most recently, Kuang and Wang further developed this technique by combining it with stimulated emission depletion (STED) microscopy.[51] STED microscopy is a type of super resolution microscopy, which can overcome the optical diffraction limit.[52,53] Figure 5.17 shows a schematic of the energy-level diagram for fluorophore. When a molecule is excited, an outer electron jumps from the ground state to the excited singlet state. From this state, de-excitation to the ground state can occur by fluorescent emission, by radiationless collision, or by crossing to the triplet state. Alternatively, return to the ground state can also be enforced by light,

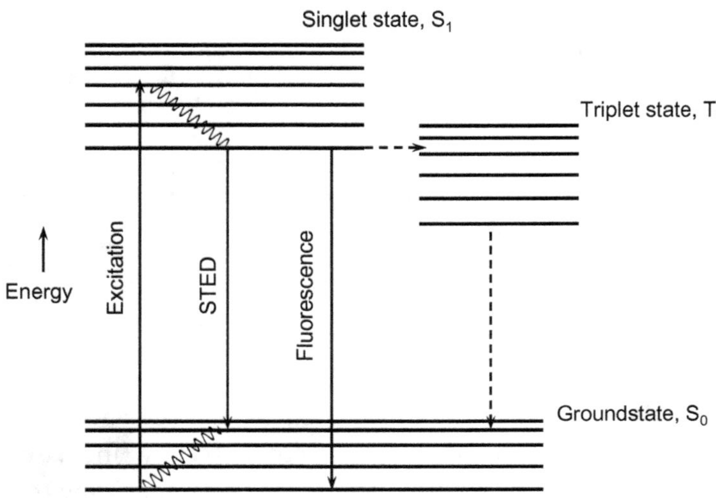

Figure 5.17. Schematic of the energy-level diagram of a fluorophore.

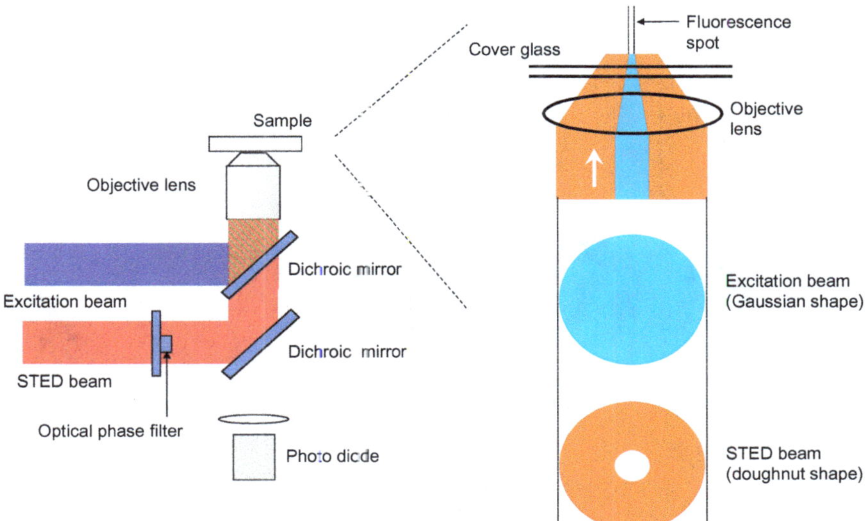

Figure 5.18. Schematic illustration of the STED microscopy. Laser beams of Gaussian shape and doughnut shape are respectively for the fluorescence excitation and STED to achieve the spatial resolution over the Abbe's diffraction limit.

through STED with the same cross-section and intensity dependence as normal absorption. Figure 5.18 illustrates a schematic of STED microscopy. The excitation laser beam, with a Gaussian intensity profile, is focused on a fluorescent sample through an objective lens and a dichroic mirror. Simultaneously, the so-called STED beam, which is doughnut-shaped by an optical phase filter, is superposed on the excitation beam, to switch off the fluorescence. Then only the sample inside the doughnut hole emits the fluorescence. Therefore, the detection spot is smaller than that of the conventional method subject to the Abbe's diffraction limit. The spatial resolution of the STED microscopy δx is described as

$$\delta x = \frac{\lambda}{2n\sin\theta\sqrt{1+I/I_{sat}}}, \tag{18}$$

where λ is the beam wavelength, n is the refractive index, q is the half aperture angle of the lens, I is the STED beam intensity, and $I_{\rm sat}$ is

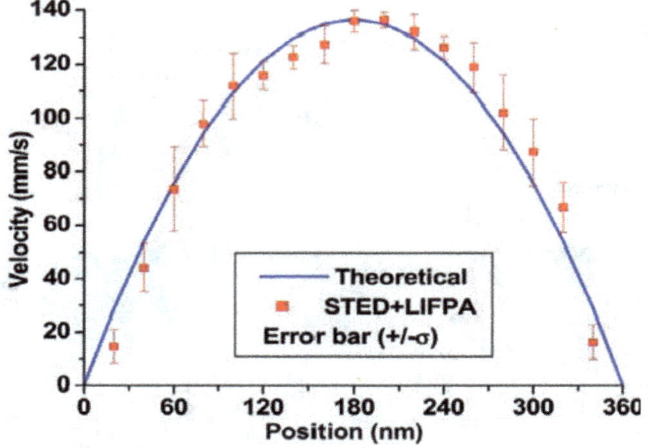

Figure 5.19. Velocity profile in a nanocapillary of 360 nm inner diameter (Adapted from Ref. 52).

the saturation intensity at which half of the excited molecules are stimulated to the ground state. Hence the detection spot can be narrowed by increasing the intensity of the STED beam. A spatial resolution of 6 nm has been realized as a maximum.[53]

A velocity profile in extended nanospace was first measured by LIFPA combined with STED. The experiment was conducted with a 360 nm nanocapillary, after the calibration between the fluorescent intensity and the velocity. Figure 5.19 shows the velocity profile of the pressure-driven flow obtained from the measurement. The result shows good agreement with a theoretical flow profile. Use of STED achieved a spatial resolution of 70 nm in the lateral direction of the laser beams. On the other hand, the resolution in the axial direction is still around 400 nm and subject to the optical diffraction limit.

References

1. Gosch M., Blom H., Holm J., Heino T., and Rigler R. (2000), Hydrodynamic flow profiling in microchannel structures by single molecule fluorescence correlation spectroscopy, *Anal Chem*, **72**, 3260–3265.

2. Le N.C.H., Yokokawa R., Dao D.V., Nguyen T.D., Wells J.C., and Sugiyama S. (2009), Versatile microfluidic total internal reflection (TIR)-based devices: application to microbeads velocity measurement and single molecule detection with upright and inverted microscope, *Lab Chip*, **9**, 244–250.
3. Bialkowski S.E. (1996), 'Chapter 1 Introduction' in *Photothermal Spectroscopy Methods for Chemical Analysis*, New York: John Wiley & Sons, Inc.
4. Tokeshi M., Uchida M., Hibara A., Sawada T., and Kitamori T. (2001), Determination of subyoctomole amounts of nonfluorescent molecules using a thermal lens microscope: subsingle-molecule determination, *Anal Chem*, **73**, 2112–2116.
5. Shimizu H., Mawatari K., and Kitamori T. (2009), Development of differential interference contrast thermal lens microscope (DIC-TLM) for sensitive individual nanoparticle detection in liquid, *Anal Chem*, **81**, 9802–9806.
6. Shimizu H., Mawatari K., and Kitamori T. (2010), Sensitive determination of concentration of nonfluorescent species in an extended-nano channel by differential interference contrast thermal lens microscope, *Anal Chem*, **82**, 7479–7484.
7. Foquet M., Korlach J., Zipfel W.R., Webb W.W., and Craighead H.G. (2006), Achieving uniform mixing in a microfluidic device: hydrodynamic focusing prior to mixing, *Anal Chem*, **76**, 1618–1626.
8. Seta N., Mawatari K., and Kitamori T. (2009), Individual nanoparticle detection in liquids by thermal lens microscopy and improvement of detection efficiency using a 1-μM microfluidic channel, *Anal Sci*, **25**, 275–278.
9. Fan F.R.F. and Bard A.J. (1997), An electrochemical coulomb staircase: detection of single electron-transfer events at nanometer electrodes, *Science*, **277**, 1791–1793.
10. Heins E.A., Siwy Z.S., Baker L.A., and Martin C.R. (2007), Detecting single porphyrin molecules in a conically shaped synthetic nanopore, *Nano Lett*, **9**, 1824–1829.
11. Rhee M. and Burns M.A. (2007), Nanopore sequencing technology: nanopore preparations, *Trends Biotechnol*, **25**, 174–181.

12. Hibara A., Saito T., Kim H.-B., Tokeshi M., Ooi T., Nakao M., and Kitamori T. (2002), Nanochannels on a fused-silica microchip and liquid properties investigation by time resolved fluorescence measurements, *Anal Chem*, **74**, 6170–6176.
13. Ren C.L. and Li D. (2004), Electroviscous effects on pressure-driven flow of dilute electrolyte solutions in small microchannels, *J Colloid Interf Sci*, **274**, 319–330.
14. Wang M., Chang C.-C., and Yang R-J. (2010), Electroviscous effects in nanofluidic channels, *J Chem Phys*, **132**, 024701-1–024701-6.
15. Bocquet L. and Barrat J.-L. (2007), Flow boundary conditions from nano- to microscales, *Soft Matter*, **3**, 685–693.
16. Neto C., Evans D.R., Bonaccurso E., Butt H.-J., and Craig V.S.J. (2005), Boundary slip in Newtonian liquid: a review of experimental studies, *Rep Prog Phys*, **68**, 2859–2897.
17. Ko H.S. and Gau C. (2009), Bonding of a complicated polymer microchannel system for study of pressurized liquid flow characteristics with the electric double effect, *J Micromech Microeng*, **19**, 115024–115037.
18. Kohl M.J., Abdel-Khalik S.I., Jeter S.M., and Sadowski D.L. (2005), An experimental investigation of microchannel flow with internal pressure measurements, *Int J Heat Mass Tran*, **48**, 1518–1533.
19. Morikawa K., Mawatari K., Kato M., Tsukahara T., and Kitamori T. (2010), Streaming potential/current measurement system for investigation of liquids confined in extended nanoscale, *Lab Chip*, **10**, 871–875.
20. van der Heyden F.H.J., Stein D., and Dekker C. (2005), Streaming currents in a single nanofluidic channel, *Phys Rev Lett*, **95**, 116104–116108.
21. Xie Y., Wang X., Xue J., Jin K., Chen L., and Wang Y. (2008), Electric energy generation in single track-etched nanopores, *Appl Phys Lett*, **93**, 163116–163119.
22. Scales P.J., Grieser F., and Healy T.W. (1992), Electrokinetics of the silica-solution interface: a flat plate streaming potential study, *Langmuir*, **8**, 965–974.
23. Oldham I.B., Young F.J., and Osterle J.F. (1963), Streaming potential in small capillaries, *J Colloid Sci*, **18**, 328–336.

24. Hunter R.J. (1981), *Zeta Potential in Colloid Science*, London: Academic Press.
25. Yang J., Lu F., Kostiuk L.W., and Kwok D.Y. (2003), Electrokinetic microchannel battery by means of electrokinetic and microfluidic phenomena, *J Micromech Microeng*, **13**, 963–970.
26. Huang X., Gordon M.J., and Zare R.N. (1988), Current-monitoring method for measuring the electroosmotic flow rate in capillary zone electrophoresis, *Anal Chem*, **60**, 1837–1838.
27. Liu Y., Fanguy J.C., Bledsoe J.M., and Henry C.S. (2000), Dynamic coating using polyelectrolyte multilayers for chemical control of electroosomotic flow in capillary electrophoresis, *Anal Chem*, **72**, 5939–5944.
28. Spehar A.M., Koster S., Linder V., Kulmala S., de Rooji N.F., Verpoorte E., Sigrist H., and Thormann W. (2003), Electrokinetic characterization of poly(dimethylsiloxane) microchannels, *Electrophoresis*, **24**, 3674–3678.
29. Sze A., Erickson D., Ren L., and Li D. (2003), Zeta-potential measurement using the Smoluchowski equation and the slope of the current-time relationship in electroosmotic flow, *J Colloid Interf Sci*, **261**, 402–410.
30. Mela P., Tas N.R., Berenschot E.J.W., van Nieuwkasteele J., and van der Berg A. (2004), Electrokinetic pumping and detection of low-volume flows in nanochannels, *Electrophoresis*, **25**, 3687–3693.
31. Pennathur S. and Santiago J.G. (2005), Electrokinetic transport in nanochannels. 2. Experiments, *Anal Chem*, 77, 6782–6789.
32. Pennathur S. and Santiago J.G. (2005), Electrokinetic transport in nanochannels. 1. Theory, *Anal Chem*, 77, 6772–6781.
33. Paul P.H., Garguilo M.G., and Rakestraw D.J. (1998), Imaging of pressure- and electrokinetically driven flows through open capillaries, *Anal Chem*, **70**, 2459–2467.
34. Maynes D. and Webb A.R. (2002), Velocity profile characterization in sub-millimeter diameter tubes using molecular tagging velocimetry, *Exp Fluids*, **32**, 3–15.
35. Mosier B.P., Molho J.I., and Santiago J.G. (2002), Photobleached-fluorescence imaging of microflows, *Exp Fluids*, **33**, 545–554.
36. Pappaert K., Biesemans J., Clicq D., Vankrunkelsven S., and Desmet G. (2005), Measurements of diffusion coefficients in 1-D micro- and nanochannels using shear-driven flows, *Lab Chip*, **5**, 1104–1110.

37. Santiago J.G., Wereley S.T., Meinhart C.D., Beebe D.J., and Adrian R.J. (1998), A particle image velocimetry system for microfluidics, *Exp Fluids*, **25**, 316–319.
38. Lindken R., Rossi M., Große S., and Westerweel J. (2009), Micro-particle image velocimetry (μPIV): recent developments, applications, and guidelines, *Lab Chip*, **9**, 2551–2567.
39. Zettner C.M. and Yoda M. (2003), Particle velocity field measurements in a near-wall flow using evanescent wave illumination, *Exp Fluids*, **34**, 115–121.
40. Li H. and Yoda M. (2008), Multilayer nano-particle image (MnPIV) in microscale Poiseuille flows, *Meas Sci Technol*, **19**, 075402–075411.
41. Li H. and Yoda M. (2010), An experimental study of slip considering the effects of non-uniform colloidal tracer distributions, *J Fluid Mech*, **662**, 269–287.
42. Meinhart C.D., Wereley S.T., and Santiago J.G. (2000), A PIV algorithm for estimating time-averaged velocity fields, *J Fluid Eng*, **122**, 285–289.
43. Sato Y., Inaba S., Hishida K., and Maeda M. (2003), Spatially averaged time-resolved particle-tracking velocimetry in microspace considering Brownian motion of submicron fluorescent particles, *Exp Fluids*, **35**, 167–177.
44. Huang P. and Breuer K.S. (2007), Direct measurement of anisotropic near-wall hindered diffusion using total internal reflection velocimetry, *Phys Rev E*, **76**, 046307–046311.
45. Guasto J.S. and Breuer K.S. (2009), High-speed quantum dot tracking and velocimetry using evanescent wave illumination, *Exp Fluids*, **47**, 1059–1066.
46. Minor M., van der Linde A.J., van Leeuwen H.P., and Lyklema J. (1997), Dynamic aspects of electrophoresis and electroosmosis: a new fast method for measuring particle mobilities, *J Colloid Interf Sci*, **189**, 370–375.
47. Vinogradova O.I., Koynov K., Best A., and Feuillebois F. (2009), Direct measurements of hydrophobic slippage using double-focus fluorescence cross-correlation, *Phys Rev Lett*, **102**, 118302–118306.
48. Pit R., Hervet H., and Léger L. (2000), Direct experimental evidence of slip in hexadecane: solid interfaces, *Phys Rev Lett*, **85**, 980–983.

49. Wang G.R. (2005), Laser induced fluorescence photobleaching anemometer for microfluidic devices, *Lab Chip*, **5**, 450–456.
50. Kuang C., Zhao W., Yang F., and Wang G. (2009), Measuring flow velocity distribution in microchannels using molecular tracers, *Microfluid Nanofluid*, 7, 509–517.
51. Kuang C. and Wang G. (2010), A novel far-field nanoscopic velocimetry for nanofluidics, *Lab Chip*, **10**, 240–245.
52. Hell S.W. (2003), Toward fluorescence nanoscopy, *Nat Biotechnol*, **21**, 1347–1355.
53. Rittweger E., Han K.Y., Irvine S.E., Eggeling C., and Hell S.W. (2009), STED microscopy reveals crystal colour centres with nanometric resolution, *Nat Photonics*, **3**, 144–147.

Chapter 6

BASIC NANOSCIENCE

6.1. Liquid Properties

6.1.1. *Introduction*

The water properties in nano-confinement geometries have been studied in various fields of chemistry, biology, and geology.[1,2] A variety of spectroscopic and theoretical investigations have studied water inside single nm-sized materials, such as carbon nanotubes, porous silica, biological macromolecules, and clay minerals. These investigations displayed unique features invoking the formation of icelike structures, slower molecular motions, and depression of the freezing point. For example, X-ray diffraction (XRD) analyses of water filled in 1 nm scale, single-walled carbon nanotubes (SWNTs) with diameters of 1.17–1.44 nm were performed, and showed that the water inside SWNTs formed icelike polygonal ring structures even at ambient temperature. The transition temperatures were found to be dependant on space size. The interactions between the water molecules and the SWNT wall could be stabilized by the icelike polygonal ring structures inside SWNTs because the SWNT wall cut out the hydrogen bonding networks of bulk water.[3] Molecular dynamic (MD) simulation studies also showed that water molecules filled in carbon nanotubes had one-dimensional ordering chain structures and, accordingly, changes in the hydrogen bonding networks, and protons of the water, could move along 1-D water chains faster than those of bulk water by a factor of 40.[4,5] A neutron diffraction study of water confined in 4 nm Vycor glass with hydrophilic surfaces indicated that the confined water was still hydrogen bonded, but the hydrogen bonding networks were strongly disordered compared with bulk water.

Namely, there are significant variations in water structures between those inside carbon nanotubes and Vycor glass.[6]

From the viewpoint of dynamics, the orientational motions of water confined in hydrophilic and hydrophobic nanoporous sol-gel glasses (2.5–10 nm) were investigated using ultrafast optical Kerr effect (OKE) spectroscopy. The results showed that the orientational dynamics of water were inhibited by the nano-confinement, and depended strongly on the surface conditions. Since the water confined in the nanoporous materials could form hydrogen bonding structures with the silanol groups of the hydrophilic pore surfaces, the orientational motions of water at hydrophilic surfaces became slower than those at hydrophobic ones.[7] One technique that can obtain the fluctuation of the hydrogen bonding network of water molecules in nano-confinement spaces is time-resolved infrared (IR) spectroscopy. OH vibrations in a 500 nm thick water layer sandwiched between thin Si_3N_4 substrates was measured, and the structural correlations in the thin water was found to be lost within 50 fs.[8]

These unique water properties inside single nm-sized materials are expected to play an important role in the appearance of chemical and biological functions, including chromatography separation of molecules, cell signaling mechanisms, and stability of protein hydration, while the materials themselves are too small to control the behavior of liquid phase molecular clusters associated with intermolecular interactions, and to be utilized as chemical analysis devices in the liquid phase.

On the other hand, an extended nanospace (10–100 nm scale space), which is located between single nanometer scale materials and microfluidic devices, has yet to be scientifically explored and is expected to be an attractive area. The extended nanospace is a transit area of molecular behavior, from an individual molecule to a bulk condensed phase, and is comparable to the thickness of an electric double layer (EDL) in an aqueous solution. In this extended nanospace, the complicated collective behavior of molecular clusters of 10–100 nm scale in a liquid phase should be characterized, and the charged surfaces make it possible to induce unusual liquid properties and/or chemical reactions that are quite different from bulk. Herein, we describe various physicochemical phenomena in one-dimensional extended nanospaces with 10–100 nm scale depth and μm width, and in two-dimensional extended nanospaces with both width and depth dimensions on the 10–100 nm scale.

6.1.2. *Viscosities of liquids confined in extended nanospaces*

The viscosities of liquids confined in extended nanospaces have been investigated by means of several systematic methods. Kitamori and coworkers measured capillary filling speeds of water into 2-D extended nanospaces having a width of 850 nm and a depth of 220 nm on a fused-silica substrate, using a high speed camera (500 frames s^{-1}) under optical microscope observation.[9] Figure 6.1 shows the procedure of capillary introduction of water into 100 μm-long extended nanospace channels, and its time course results. The front line of water was driven along the extended nanospace channels at 10 ms. When water is introduced into nanospaces due to capillary filling, the driving force is the surface energy, i.e. the Laplace pressure (ΔP) is described by the following equation:

$$\Delta P = \frac{2\gamma}{r}, \tag{1}$$

where γ and r denote the surface tension (73 dyn cm^{-1} for water) and the radius of curvature, respectively. In the case of rectangular shaped 2-D extended nanospace channels with width (W) 330 nm and depth (D) 220 nm, the rectangular shaped space size can be replaced by the cylindrically shaped space size with equivalent diameter ($R = 264$ nm) according to $4/R = 2(D + W)/DW$. Since the r corresponds to half of R, ΔP can be estimated as about 1.0 MPa. The pressure drop (ΔP_d) in the fluid flows is generally expressed as follows;

$$\Delta P_d = \frac{32\eta v L}{R^2}, \tag{2}$$

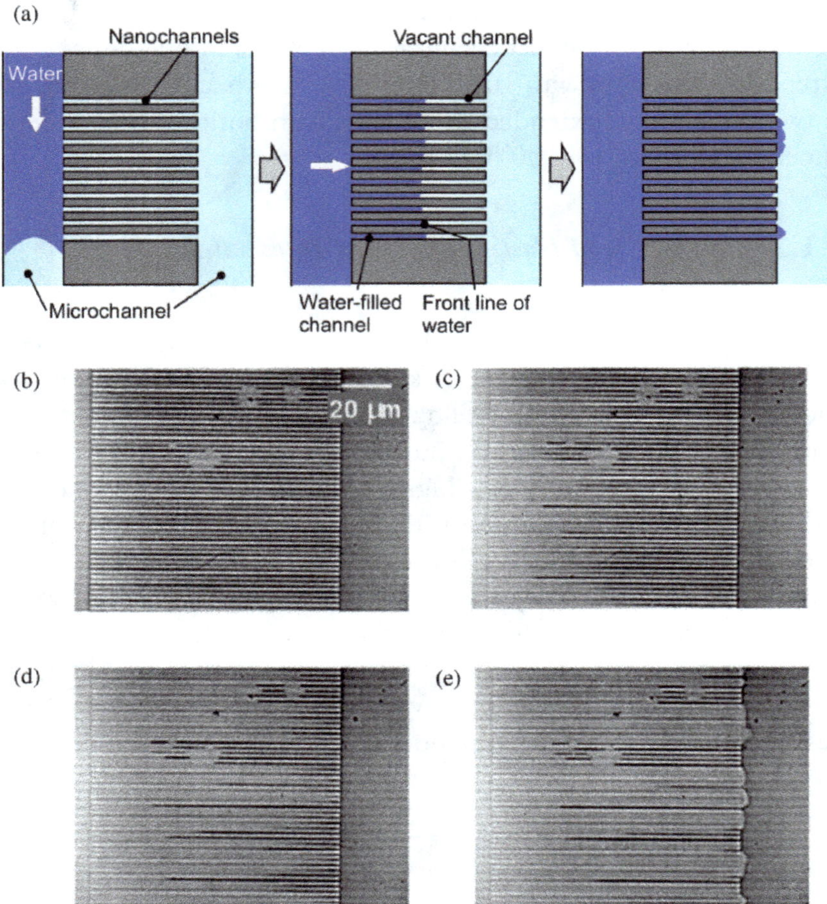

Figure 6.1. (a) Experimental procedures of capillary filling of water into extended nanospace channels; after the water was filled in microchannels, the water could be introduced into extended nanospace channels from the left side to the right side by capillary forces. Optical microscope images of the capillary filling phenomena of water into extended nanospace channels at (b) 0 ms, (c) 4 ms, (d) 8 ms, and (e) 12 ms (Redrawn from Ref. 9).

where η, v, and L represent the viscosity of water, the velocity of water, and the nanospace channel distance between two points, respectively. If we assume that the ΔP_d is comparable to ΔP, we can determine the theoretical water introduction time $\langle t \rangle$ into the extended nanospace channel length (L' = 100 μm) based on the following equation

$$\langle t\rangle = \int_0^{L'} \frac{dL}{v} = \int_0^{L'} \frac{32\eta L}{R^2 \Delta P} dL = \frac{16\eta L'^2}{R^2 \Delta P}. \tag{3}$$

From these assumptions, the introduction time, $\langle t\rangle$, can be determined as 2.3 ms. This time was found to be much faster than the experimental time of 10.0 ms. This discrepancy suggests that the filling speed of water decreased with a decrease in space sizes, and that the water in the extended nanospace channels had a viscosity about 3 times higher compared with bulk water.

The capillary filling speeds of hydrogen-bonded solvents, such as water and alcohol, were also measured by van der Berg *et al.* for 1-D extended nanospaces with a height down to 50 nm, by Staufer *et al.* for 2-D SiO_2/Si_3N_4 extended nanospaces with a width of 900 nm and depth of 27–73 nm, and for even smaller cross-sections down to sub-10 nm by Elwenspoek *et al.*[10–12] The Washburn equation (equation (1)), which is related to the moving distance, x, of the meniscus, the surface tension γ, the viscosity η, contact angle θ, channel height h_0, and filling time t, respectively, makes it possible to theoretically discuss the capillary filling kinetics:

$$x = \sqrt{\frac{\gamma \cos\theta h_0}{3\eta}}\sqrt{t}. \tag{4}$$

The comparison between experimental data and the theoretical data based on the Washburn equation showed that the filling speeds of water decreased with a decrease in space sizes because of the increase in water viscosity due to nano-confinement. Such viscosity enhancement phenomena in extended nanospaces on a glass substrate, could be explained by the formation of hydrogen bonding between liquid and the silanol groups of hydrophilic glass surfaces.

When the hydrophilic surface is converted to a hydrophobic surface, the filling speed of liquids is not quite consistent with the hydrophilic case. Quirke and coworkers recently demonstrated pressure-driven flow experiments of water or ethanol into extended nanoscale carbon nanotubes (~50 nm), and found that their filling speeds into the hydrophobic nanotube channels could be more than 10 times faster than the theoretical values predicted from the Hagen–Poiseuille equation.[13] It is likely that the filling speed enhancement phenomenon is related to either the viscosity drop of the water inside the hydrophobic nanotube channels, or the water–nanotube surface intermolecular interactions.[13,14]

Moreover, the time-resolved fluorescence measurements of aqueous solutions containing rhodamine fluorescent dyes, confined in 2-D extended nanospace channels with width 330 nm and depth 220 nm and on a 850 nm wide and 220 nm deep fused-silica chip were performed, and the size-dependence of the viscosity of water was evaluated. It was found that the fluorescent decay times of the fluorescent dyes in the solutions were dependent on the size of the channel, while there were not any differences in the fluorescence spectra regardless of the nanospace channel sizes. These results indicated that water confined in 2-D extended nanospace channels has a viscosity a few times higher and a dielectric constant (polarity) a few times lower than bulk water.[9,14] The hydrodynamic flow of water molecules in extended nanospace pillars (height: 400 nm, spacing: 500 nm) was examined by using single-particle tracking (SPT) methods, and the diffusion coefficient of 50 nm diameter nanospheres in water was found to be smaller than the theoretical value predicted by the Stokes–Einstein model. The authors concluded that the viscosity of water confined in the nanospaces is 3 times higher than that of bulk water.[16]

6.1.3. *Electrical conductivity in extended nanospaces*

Ion transport through extended nanospaces can be estimated by the ionic strength dependence of the conductance value for the electrolyte solution. Electric conductance of electrolyte solutions in nano-confinement spaces has been investigated using impedance spectrometry. Renaud and coworkers realized AC impedance measurements of KCl solutions filled in 1-D extended nanospace channels with 50 nm height, using platinum electrodes embedded in microchannels, interfaced with the nanospace channels.[17,18] Since the resulting Cole–Cole plots were both semi-circular at high frequency and approached the Warburg impedance line at low frequency, it was possible to obtain the solution resistance, R, in the extended nanospace, at each individual KCl concentration, from the intersection of the semi-circle with the real axis. When the conductance, G, values, which could be described as the reciprocal of R, were plotted against the KCl concentrations in the nanospaces, the G values were found to decrease with decreasing KCl concentrations, reaching a conductance plateau for low ionic concentrations below ~10^{-4} M. This conductance plateau could be explained by the fact that the Debye length was comparable to the extended nanospace channel height and that mobile counter-ions, i.e. excess protons, were enriched in the extended nanospaces relative to the number of protons expected from a given bulk solution, since the magnitude of the conductance plateau shifted to lower values with a decrease in the density of ionized silanols (SiO^-) on the surface.

Kitamori and coworkers constructed a novel nanofluidic chip equipped with 2-D extended nanospaces and two microchannels containing mercury (Hg) microelectrodes, as shown in Figure 6.2(a), and tried to characterize the confinement-induced nanospatial properties from the conductance data of KCl solutions in the extended nanospaces.[19] The nanofluidic chip can only detect impedance data in extended nanospaces, excluding bulk information, because the Cole–Cole plot obtained could be a simple semi-circle at high frequency, without invoking Warburg impedance (see Figure 6.2(b)). Since the real axis of a semi-circle, and the top of the semi-circle on the Cole–Cole plot, correspond to the solution resistance R and frequency f, respectively, the specific resistance, ρ, and the capacitance, C, values of the solutions can be determined according to $\rho = RA/d$ and $RC = 1/2\pi f$. Here, d/A is the cell constant. As seen from Figure 6.2(c), although the ρ values were found to increase gradually with a reduction in KCl ion concentrations, the concentration dependence of C values was opposite to the tendency of ρ. These results indicated that the viscosity, η, and the dielectric constant, ε, of KCl electrolyte

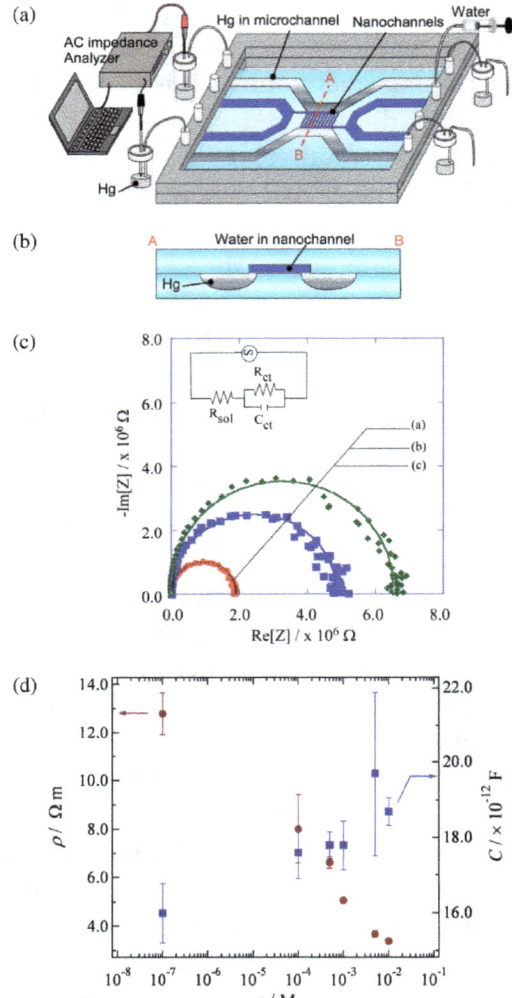

Figure 6.2. (a) Schematic illustration of the setup for a nanofluidic electrical impedance measurement system equipped with liquid Hg electrodes and extended nanospace channels on a glass chip, (b) A cross section from A to B of the nanofluidic chip, (c) Typical Cole–Cole plots of several KCl solutions ((a) 1.0×10^{-7} (b) 1.0×10^{-4} (c) 1.0×10^{-2} [M]) in 216 nm-sized extended nanospaces at 22 °C. The inset shows the equivalent circuit that consists of: R_{ct} is resistance of solutions including the effects of charged surface (interface solutions), C_{ct} is the capacitance of interface solutions, and R_{sol} is resistance of bulk electrolyte solutions, (d) Semi-logarithmic plots of ρ (●) and C values (■) *vs.* concentration of KCl (Redrawn from Ref. 19).

solutions were increased and decreased by nanospace confinement, respectively, because the ρ and C values are roughly related to the η and ε values. Moreover, the conductance, G, of the sample was obtained as the reciprocal to R. The plots of G values vs KCl concentrations in a 2-D nanospace showed a conductance plateau below ~10^{-4} M. This tendency was consistent with the results from 1-D nanospaces. The experimental G value at 1.0×10^{-7} M was obtained for $\sigma = 0.04$ C m^{-2}, which was about one-half lower than the theoretically calculated G values, with a surface charge density of $\sigma = 0.07$. From these results, it was confirmed that, in the extended nanospaces, the density of ionized silanols (SiO^-) on a surface was reduced and that proton concentration was greater than the number of protons expected from a given initial solution.

The conductivity measurements of the $HClO_4$ solution filled in various 1-D extended nanospace channels made clear that proton conductivity increased with the reduction in channel depth below about 1 μm. In particular, the proton conductivity of 10^{-3} M $HClO_4$ solution in a 50 nm deep nanospace channel became greater than that in a given bulk solution by a factor of 30. The authors concluded that an enhancement of proton conductivity was caused by the EDL overlap in the nanoconfinement spaces.[20]

6.1.4. *Streaming current/potential in extended nanospaces*

Since the EDL induces an electric potential gradient in the extended nanospaces, electrochemical experiments can be employed to evaluate the effects of surface charge on liquid properties. In particular, when the counter-ions in the EDL are transported in the nanospace channels by pressure-driven hydrodynamic flow, a streaming potential/current can be induced. Since the detailed information about the measurement principles is available in Section 5.2, only unique properties are explained in this section. The streaming currents of electrolyte solutions containing various ions, such as K^+, Ca^{2+}, and Mg^{2+}, in 1-D silica nanospace channels with 10–1000 nm-sized channel depth were measured by van der Heyden and coworkers.[21,22] The streaming currents, which are induced by the nano-confinement, were found to be dependent on the charge density, the flow rate in the fluids, and the ion strength of the fluids. Moreover, the authors clarified that the pressure-driven transport energy of counter-ions inside the 1-D nanospace channels made it possible to convert to electrical power with an efficiency of a few per cent.[23,24] Such phenomena could be explained by the combination of Poisson–Boltzmann theory, with the effects of EDL overlap and surface charge density for electrical fields, and Hagen–Poiseuille law for flows (see Section 5 for detailed principles). Since the electrostatic potential distribution $\psi(r)$ and the fluid viscosity $u(r)$ at the distance r are related to the electrical resistance of the nanospace channel, the fluidic impedance of the nanospace channel, the streaming conductance, and the energy conversion efficiency can be theoretically evaluated for the nanospace channels. Although the experiments streaming currents of KCl solutions above a concentration of 10^{-4} M were consistent with theoretical values, a discrepancy between experimental and theoretical values appeared below 10^{-4} M. Such a discrepancy could well be explained by the fact that changes in net surface charge density resulted from the dissociation of protons from SiOH groups on the nanochannel surfaces.

A novel streaming potential/current measurement system for 2-D extended nanospaces with more than 200 nm was developed by Kitamori and coworkers.[25] The dependence of pressures, channel surface areas, and surface charges upon streaming potential/current were examined in various 2-D extended nanospaces. These results showed that the flow rates in the extended nanospace channels increased with an increase in the applied pressures, because the potential/current values were proportional to the applied pressures. The authors demonstrated that the electric signals obtained from the nanospace channels were quite sensitive to the surface charge conditions. They demonstrated that positive voltages and currents could be obtained in the non-modified extended nanospace channel with negatively charged silanol surface because of cation flows inside the nanospace channel, while the positively charged amino groups on a 3-aminopropyltriethoxysilane (APTES)-modified surface could generate the negative voltages and currents due to anion flows inside the APTES nanospace channel.

6.1.5. *Ion transport in extended nanospaces*

Several researchers have demonstrated that unique ion or molecular transport phenomena can be generated by EDLs in extended nanospaces. When cations and anions are introduced into extended nanospaces fabricated on a negative silica glass surface, the cations will be enriched near the negative silica surface and the anions localized to the center of the nanospaces. In such uneven ion distributions, anions should flow faster than cations, as the flow profile in the nanospaces is parabolic. Therefore, Kato and coworkers investigated the effect of EDL thickness on the velocity of fluorescent solutes in the 2-D extended nanospace channels (width; 270 nm, depth; 280 nm) using pressure-driven flows.[26] The results showed that the EDL thickness had a large effect on the velocity of the solutes, and that the more negatively charged solutes such as fluorescein, which is negative and bivalent, produced a higher velocity compared with sulforhodamine B, which is monovalent, or rhodamine B, which is neutral, as shown in Figure 6.3. This is evidence that negatively charged ions could be localized to the center of the extended nanospaces, because the Debye length, about 50 nm for a solution containing 50 μM fluorescent dyes, approached a size comparable to that of the used extended nanospace.

Ion enrichment/depletion in 1-D extended nanospace channels was further demonstrated using the transport phenomena of fluorescent dye molecules. When a voltage was applied to a solution containing negatively charged fluorescein dye and positively charged rohdamine dye and that solution filled in the 1-D extended nanospace channel, both the fluorescein and rohdamine dyes were enriched at the cathode end and depleted at the anode end of the nanospaces.[27]

6.1.6. *Gas/liquid phase transition phenomena in extended nanospaces*

In confining geometries, the saturated vapor pressure is generally lower than that at a flat surface due to Laplace pressure. Namely, when a contact angle of the corresponding liquid, θ, is less than 90°,

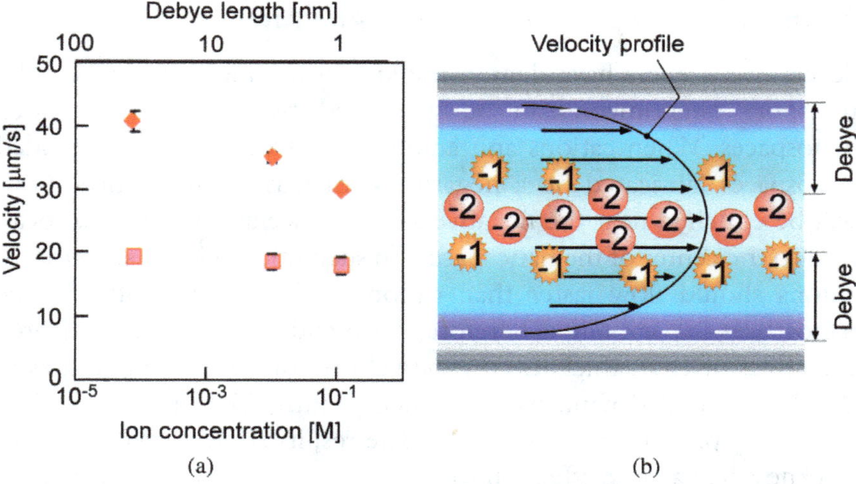

Figure 6.3. (a) Effects of ionic concentration (Debye length) on the velocities of solutes with different charges. The symbols ♦ (red) and ■ (pink) represent fluorescein (−2) and sulforhodamine B (−1), respectively, (b) Schematic illustration of the distribution of solutes and their velocity profile in the extended nanospace channels (Redrawn from Ref. 26).

the saturated vapor pressure in the confined nanospace p_r is less than the saturated pressure on a flat surface p_0 according to the following Kelvin equation:

$$\ln \frac{p_r}{p_0} = -\frac{2V_m \gamma \cos\theta}{RTr}, \tag{5}$$

where V_m, γ, and r are molar volume, surface tension, and capillary radius of the nanospaces, respectively.

However, there is not any experimental evidence regarding the applicability of the Kelvin equation in extended nanospaces. Kitamori *et al.* constructed a vapor–liquid phase transition measurement apparatus equipped with micro- and 2-D extended nanospace channels on a chip. In this system, the temperatures and humidity of vapor in the nanospace channels could be strictly controlled. They examined the size- and time-dependence of capillary evaporation phenomena of water using optical microscope observation. Notably, the water was found to be evaporated in microchannels at 22.0 °C, while it did not evaporate in 2-D extended nanospaces under the same temperature. When the temperature reached 22.2 °C, the water in 120 nm extended nanospaces evaporated gradually. The experimental vapor pressure values of water in 2-D extended nanospaces were recorded as lower than those calculated from a model based on Kelvin's equation.[28]

By utilizing the lower vapor pressures in extended nanospaces, Kitamori and coworkers demonstrated distillation of an aqueous solution containing 9.0 wt% ethanol, and capillary condensation of ethanol vapor in extended nanoscale pillar structures by observing the optical diffraction patterns during the condensation process.[29] As shown in Figure 6.4, the diffraction patterns of ethanol vapor at 270 nm-sized extended nanopillars, which were arranged between two microchannels on a glass chip, were observed before and after capillary condensation. Although purple diffraction light was observed in the initially vacant nanopillars, the purple area turned gray once the condensed ethanol filled the vacant nanopillars, even at the boiling point of 78.0 °C. This capillary evaporation/condensation phenomena in extended nanospaces apparently indicated that the existence of liquid-charged surface interactions plays an important role for determining the saturated vapor pressure of liquids.

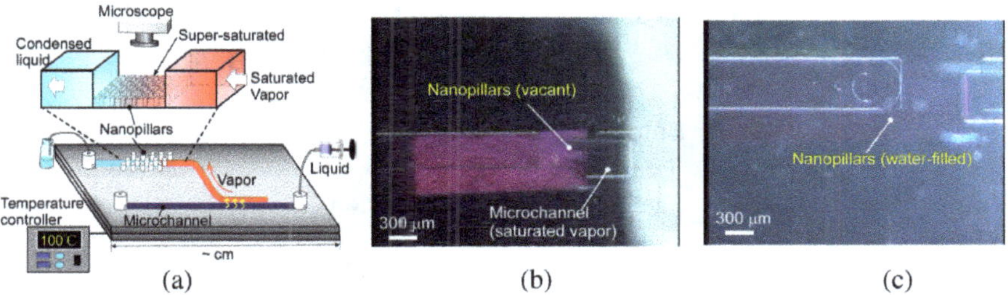

Figure 6.4. (a) A schematic illustration of capillary condensation measurement in extended nanospace channels. When the liquid was filled in a microchannel with deep width on a heated chip, only vapor could be introduced into nanopillars embedded with a microchannel having shallow width. Micrographs of the nanopillars (b) before and (c) after capillary condensation (Redrawn from Ref. 29).

6.1.7. *Structures and dynamics of liquids confined in extended nanospaces*

Limited information is available at the molecular level about how nanospace confinement affects liquid properties, and to what extent the liquid properties in extended nanospaces are different from those in bulk or single nm-sized materials. A nuclear magnetic resonance (NMR) method, which can detect the variation of molecular structures and dynamics as listed in Table 6.1, is suitable for characterizing the complicated behavior of liquid molecules confined in extended nanospaces. Kitamori and coworkers fabricated an NMR cell equipped with 40–5000 nm extended nanospaces, and employed the H_2O, ^2H_2O, and $H_2{}^{17}O$ results of a ^{1}H-NMR chemical shift (δ_H) and ^{1}H- and ^{2}H-NMR spin-lattice relaxation rate (^1H- and ^2H-$1/T_1$), ^{1}H-NMR spin-spin relaxation rate (^1H-$1/T_2$), and ^{1}H-NMR rotating-frame spin-lattice relaxation rate (^1H-$1/T_{1\rho}$) to clarify confinement-induced

Table 6.1. Relationship between various NMR relaxation rate measurements and obtained information for examining molecular dynamics.

	Spin-lattice relaxation rate ($1/T_1$)	Spin-spin relaxation rate ($1/T_2$)	Rotating-frame spin-lattice relaxation rate ($1/T_{1\rho}$)
Relaxation processes	After excitation External field Before excitation Spin	Coherent Random	B$_1$ (Rocking field) B$_1$
Method	Inversion recovery	CPMG	Spin-locking
Frequency resion	High freq. (~MHz)	High + Low	Low freq. (~KHz)
Information	Motions (Trans.lRot.)	• Motions • Proton transfer	• Proton transfer • Magnitude of Dipole-dipole interaction

nanospatial properties of molecular structures and dynamics of water and non-aqueous solvents.[30,31]

The δ_H results showed that the water and non-aqueous solvents confined in 40–5000 nm extended nanospaces retain the four-coordinated hydrogen-bond structures without changing the O–O distance between H_2O molecules, as seen for ordinary liquid H_2O, because their δ_H values were almost constant regardless of space sizes. On the other hand, the $^1H\text{-}1/T_1$ values for water confined in extended nanospaces with decreasing sizes showed marked variation, as shown in Figure 6.5. Although the $^1H\text{-}1/T_1$ values were almost constant for space sizes of 800–5000 nm, the values changed drastically below about 800 nm, increasing continuously with decreasing space sizes up to around 200 nm. Below 200 nm, size-dependence of

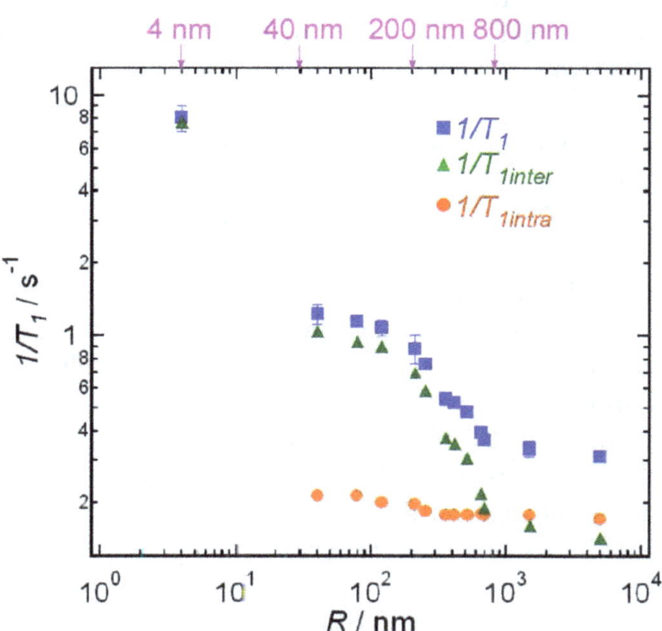

Figure 6.5. Log-Log plot of the measured *1H-1/T1* values (■), intermolecular translational component ($1/T_{1inter}$) (▲), and intramolecular rotational one ($1/T_{1intra}$) (●) for water confined in $R = 4$–5,000 nm spaces at 500 MHz and 22.0 °C (Adapted from Ref. 30).

the $^1H\text{-}1/T_1$ values disappeared. When the size was further reduced to 4 nm controlled porous glass, an increase in $^1H\text{-}1/T_1$ values was again observed. Generally, the $^1H\text{-}1/T_1$ values can be accounted for by the contributions of intermolecular translational motion $(1/T_{1inter})$ and intramolecular rotational motion $(1/T_{1intra})$ as follows:

$$\frac{1}{T_1} = \left(\frac{1}{T_{1intra}}\right) + \left(\frac{1}{T_{1iner}}\right) = \frac{3}{2}\frac{\gamma^4\hbar^2}{r^6}\tau_R + \frac{\pi}{5}\frac{N\gamma^4\hbar^2}{aD}, \tag{6}$$

where $\hbar$ is the Plank constant, r is the distance between the hydrogen atoms in a H_2O, γ is the proton gyromagnetic ratio, a is the hydrodynamic radius, N is the number of spins per unit volume, k_B is the Boltzmann's constant, T is the temperature, D is the proton's self-diffusion coefficient, and τ_R is the molecular reorientational correlation time. Here, τ_R is related to $4\pi\alpha^3\eta\ /3\ k_BT$ with viscosity η through the Stokes–Einstein–Debye hydrodynamic equation.

Since the $^2H\text{-}1/T_1$ values of heavy water (^2H_2O) associated with only rotational motion components did not depend on the variation of space sizes, the $^1H\text{-}1/T_1$ results suggested that only the translational motions of H_2O were inhibited by size confinement.

The $^1H\text{-}1/T_1$ values provide insight into the faster motions involving the higher frequency component, whereas the $^1H\text{-}1/T_2$ and $^1H\text{-}1/T_{1\rho}$ values are sensitive to the slower motions related to the lower frequency component. The $1/T_2 - 1/T_1$ values of water increased more than two orders of magnitude by changes from 5000–40 nm, and the differences approached the value for adsorbed water. Such enhancement of the $1/T_2 - 1/T_1$ values can be roughly associated with changes in the proton exchange rates in H_2O molecules. Since the specific interface area of the glass surfaces becomes very high in the extended nanospaces, proton hopping between water and charged surfaces was induced along the linear O···H–O hydrogen bonding chains, i.e. (°SiOH + H_2O) + H_2O → ≡SiO^- + (H_3O^+ + H_2O) → ≡SiO^- + (H_2O + H_3O^+). This results in the enhancement of proton transfer between H_2O molecules. The $^1H\text{-}1/T_{1\rho}$ values led to the conclusion that the proton transfer rate of water in the 100 nm-sized extended nanospaces could be larger by a factor of more than ten compared with that of bulk water. Such phenomena were characteristics common to hydrogen-bond solvents, such as alcohols, but did not appear in aprotic and non-polar solvents, such as hexane.

Based on these NMR results, Kitamori *et al.* hypothesized that a proton transfer phase, S_P, in which the H_2O molecules are loosely coupled within about 50 nm via hydrogen bonds in a direction perpendicular to the glass surfaces, exists mainly in extended nanospaces. The existence of such long-range structuring of hydrogen-bonded solvents on a surface has been recently suggested from the results of MD simulations.[32] The validity of this suggestion was supported by a three-phase theory where the water behavior in nano-confining geometries was controlled by the weighted average of three phases, such as the bulk phase S_B, with ordinary water structure and free translational and rotational motions, the adsorbed phase S_a, with icelike bilayer structure and slower translational and rotational motions, and the S_P, as illustrated in Figure 6.6.

6.2. Chemical Reaction

Since unique liquid properties in extended nanospaces are expected to affect chemical reactions, there will be differences apparent in reaction properties between the nanospaces, bulk spaces, and microspaces. Thus, it is essential to evaluate precisely which chemical reaction mechanisms and molecular transport phenomena are involved. Taking this viewpoint, several important chemical reactions have been demonstrated in extended nanospaces as mentioned below.

6.2.1. *Enzymatic reaction*

Kitamori and coworkers developed a pressure-driven nanofluidic control system and realized an enzyme reaction, in which the fluorogenic substrate TokyoGreen-β-galactoside (TG-β-gal) was hydrolyzed to fluorescein derivative TokyoGreen (TG) and β-galactose, by β-galactosidase enzymes acting as a catalyst in a Y-shaped extended nanospace channel, as shown in Figure 6.7(a).[33] They evaluated size-confinement effects

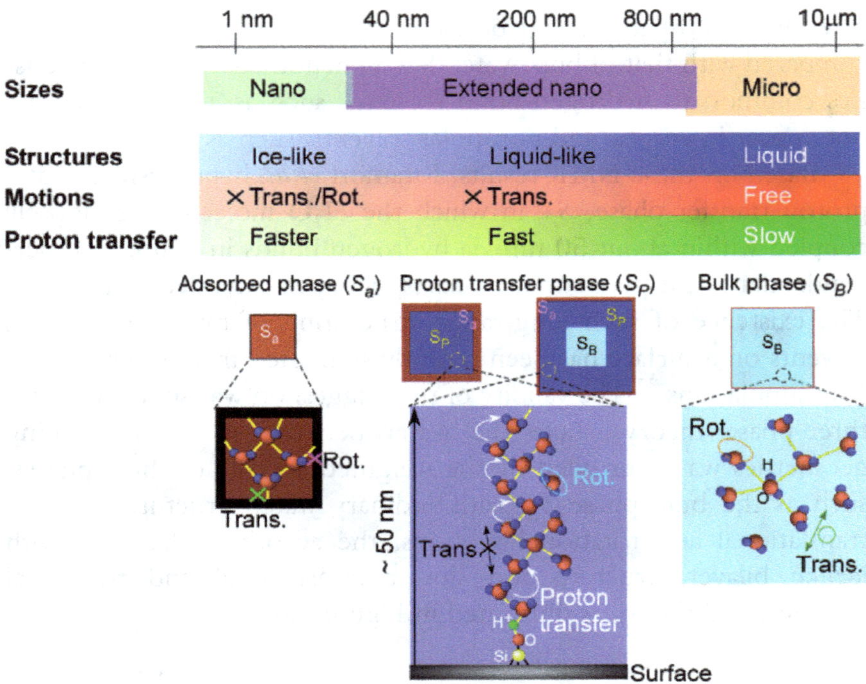

Figure 6.6. Schematic pictures illustrating the three-phase model: In S_B, the water molecules have an ordinary liquid structure and free translation and rotation. In S_P, the water molecules have the properties such as keeping four-coordinated H_2O structure, slower translational motion, and higher proton transfer due to proton hopping along a linear O···H-O hydrogen bonding chain. In S_a, the water is similar to an icelike bilayer structure and both translation and rotation are inhibited.

on the kinetic parameters for the enzyme reaction, such as the Michaelis constant, K_m, the maximum reaction rate, V_{max}, and the first-order rate constant of the enzyme reaction, k_{cat}. In particular, the k_{cat} value for the nanofluidic reaction was compared with the results of the bulk and microfluidic reactions, and their comparison showed that the enzyme reaction rate for the nanofluidic system increased by a factor of approximately two when compared with those for bulk and microfluidic systems (Figure 6.7(b)). The acceleration of the reaction kinetics is attributable to an enhancement of the proton mobility of water in the extended nanospaces.

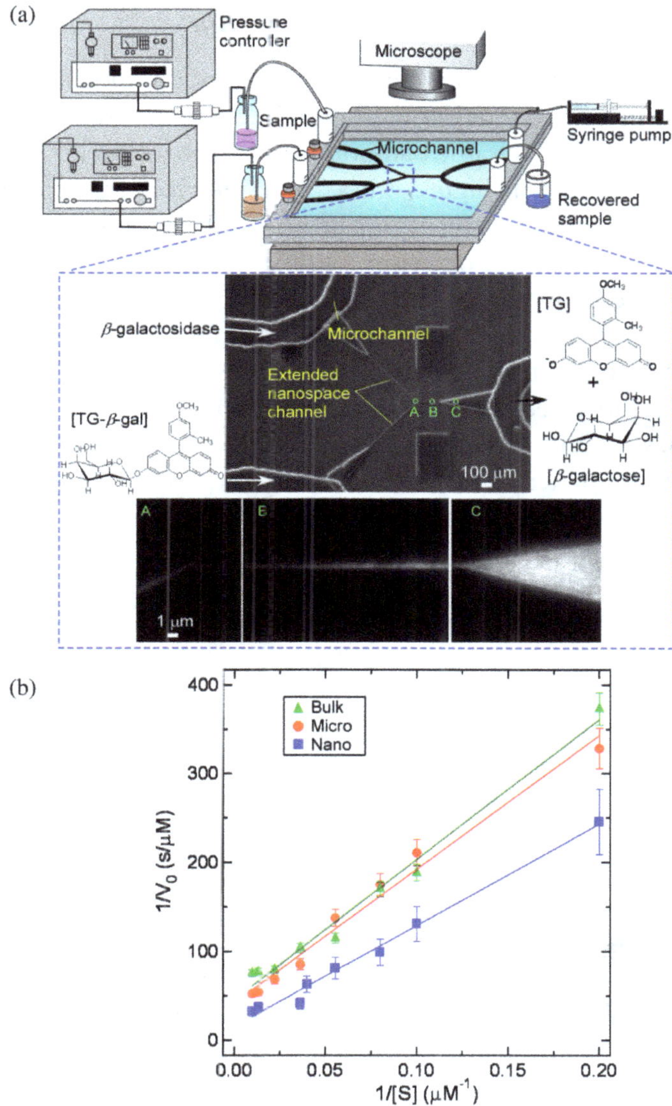

Figure 6.7. (a) Schematic illustration of nanofluidic control system, and fluorescence micrograph of an enzymatic reaction inside a Y-shaped extended nanospace channel, (b) Lineweaver–Burk plots of the initial enzymatic reaction rate [V_0] *vs.* TG-b-gal substrate concentration [S] in bulk, microfluidic space (100 μm width and 40 μm deep), and extended nanospace channel (520 nm width and 234 nm deep). The k_{cat} values were determined from the slope of the line and the intercept (Redrawn from Ref. 33).

6.2.2. Keto-enol tautomeric equilibrium

Keto-enol tautomerization refers to the one of the most basic chemical equilibrium reactions between two different structures (keto form and enol form) of the same compound. The Hacac has keto-enol tautomers as depicted in Figure 6.8(a). The interconversion of the two forms involves the movement of a proton and the shifting of bonding electrons; hence, the isomerism qualifies as tautomerism. Kitamori and coworkers investigated the keto-enol tautomeric equilibrium of acetylacetone (Hacac) in water inside 2-D extended nanospaces based on ^{1}H-NMR spectra analyses.[34] The ^{1}H-NMR peaks of Hacac in water were observed for a size range from 202–5000 nm, as shown in Figure 6.8(b), and the size-dependence of peak area ratios due to the $-CH_3$ groups of keto and enol forms (K_{EQ} = [keto]/[enol]) of Hacac

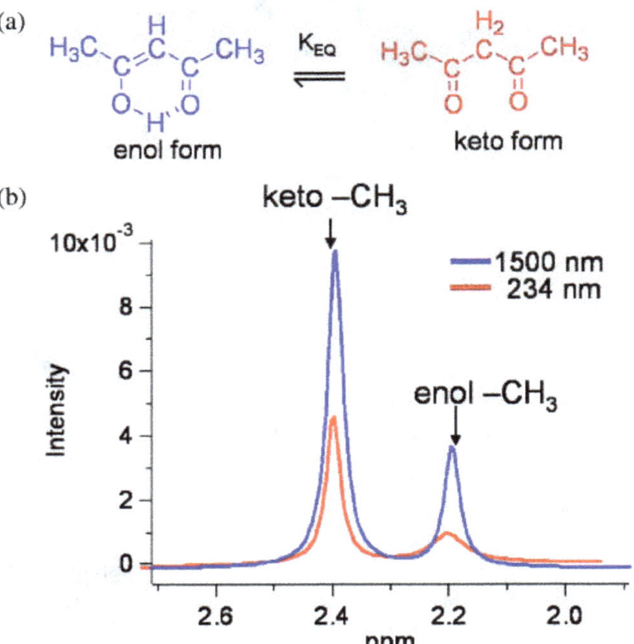

Figure 6.8. ^{1}H-NMR spectra of –CH3 groups of keto and enol forms for Hacac solutions confined in 234 nm and 1,500 nm at 20 °C.

were examined. Noteworthily, the K_{EQ} values were found to increase drastically with decreasing sizes at around 500 nm, i.e. enhancement of the keto form of Hacac by size-confinement. The K_{EQ} values for the bulk and 202 nm are 3.7 and 4.8, respectively. This tendency of K_{EQ} could be associated with the fast proton chemical exchange between Hacac and water, which induces the shift to a keto form tautomer with decreasing size below 500 nm, because the K_{EQ} values were reduced by deuterium (D) substitution of water (from H_2O to $H_2O + D_2O$ mixture).

6.2.3. *Nanoparticle synthesis*

Colloidal particles or droplets with controlled sizes and morphologies have been synthesized inside the 1-D extended nanospace channels with a depth of 420–1000 nm and U-shaped microchannels.[35] By using pressure-driven nanofluidic flows, water and oil solutions contained in the different microchannels were introduced into a cross-shaped 1-D extended nanospace channel, and the immiscible solutions produced a sheath flow at the cross junction. After the sheath flow was released downstream of the nanospace channel into a microspace chamber, nm-sized droplets formed due to the interfacial tensions between the immiscible solutions, which were strongly influenced by the size and structure of the nanochannel. This nanofluidic device makes it possible to synthesize simple and multiple droplets with a minimum diameter as small as 900 nm.

6.2.4. *Nano DNA hybridization*

Patel *et al.* demonstrated the electrokinetic DNA hybridization in a T-shaped 1-D extended nanospace channel with width 10 μm and depth 200 nm.[36] Two solutions, without gel, could be diffusion-mixed at a T-intersection in the extended nanospace channel using an electric field of 220 Vcm^{-1}. This would cause almost all the single-stranded DNA (ssDNA) fragments to be converted into double-stranded DNA (dsDNA) fragments at a hybridization time of about 600 s. It is most likely that this nano-hybridization is a highly efficient method, because there is no need to provide either long hybridization time or sample labeling and sieving structures.

6.2.5. *Nano redox reaction*

A nanofluidic device, in which two platinum electrodes and a 1-D extended nanospace channel with a height of around 50 nm, embedded in a SiO_2 substrate, were fabricated using nanofabrication techniques, and the chemically reversible redox reaction of a ferrocenedimethanol ($Fc(MeOH)_2$) solution confined in the extended nanospaces was demonstrated using cyclic voltammetry.[37] The results showed that the heterogeneous electron-transfer rate constant of $Fc(MeOH)_2$ was quite fast and changed with the addition of supporting electrolytes. The rate constant was expected to depend on the structure of the electrical double layer formed in the extended nanospace, because the double layer can become comparable to the size of the diffusion layer.

6.3. Liquid Properties in Intercellular Space

Here, we introduce a report on the creation of biomimetic space using extended nanotechnology, shown in Figure 6.9.[38] Many researchers have shown that nanometer spaces between cell membranes and synaptic gaps of order 10 nm, play a very important role in cell functions (signaling, tissue formation, etc.), and have suggested specific solution properties close to cell membranes.[2] It is therefore necessary to make a tool for investigating solution properties between them, but there is, as yet, no experimental tool. Conversely, as described in the previous sections, the extended nanospace in glass substrates was recently studied, and many specific liquid properties have been found. Properties of water such as viscosity, dielectric constant, and proton transfer, for example, are different in the extended nanospace from those in the bulk space.[30,31] It has been suggested that there are similar properties between the extended nanospace and intercellular space. As such, we are interested in producing biomimetic extended nanospaces containing cell membranes. To realize this space, we needed to modify phospholipid bilayers, which are the

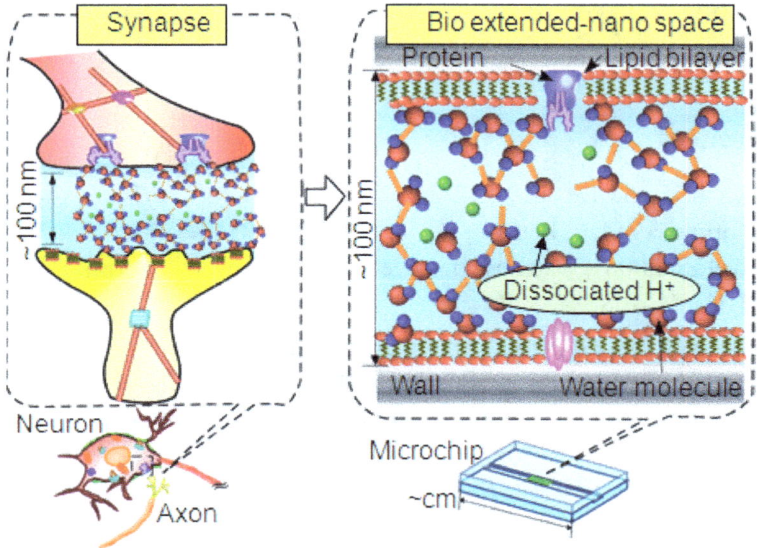

Figure 6.9. Concept of biomimetic extended nanospaces.

main component along with cell membranes, in this space. In this report, conditions for modifying phospholipid bilayers in the extended nanospace were investigated.

The extended nanospace channels (of equivalent diameter R = 390–620 nm) were fabricated on a synthetic quartz glass by EB lithography and plasma etching. Then, microchannels and holes for sample and mobile phase introduction were fabricated on the chip. The chip was thermally bonded to an upper plate at a temperature of 1080 °C. As phospholipids, 99% 1,2-Dioleoyl-sn-Glycero-3-Phosphocholine (DOPC), including 1% dye-conjugated lipids (Texas-Red DHPE), were used. The buffer solution was filled into the chip and the vesicle solution was flown in the extended nanochannels for 5–10 minutes. For washing the suspended vesicles, the buffer solution was flown in the chip again and the modified phospholipid bilayers were observed through fluorescent microscopy. As a result, phospholipid bilayers were actually modified in the extended nanospace. In addition, the solution flow in nanochannels was checked by watching tiny aggregated vesicles. This system mimicking cell–cell space can be applied for investigating water properties between cells.

References

1. Buch V. and Devlin J.P., Eds. (2003), *Water in Confining Geometries*, Springer-Verlag, Berlin.
2. Ball P. (2008), Water as an active constituent in cell biology, *Chem Rev*, **108**, 74–108.
3. Maniwa Y., Kataura H., Abe M., Udaka A., Suzuki S., Achiba Y., Kira H., Matsuda K., Kadowaki H., and Okabe Y. (2005), Ordered water inside carbon nanotubes: formation of pentagonal to octagonal ice-nanotubes, *Chem Phys Lett*, **401**, 534–538.
4. Hummer G., Rasaiah J.C., and Noworyta J.P. (2001), Water conduction through the hydrophobic channel of a carbon nanotube, *Nature*, **414**, 188–190.
5. Dellago C., Naor M.M., and Hummer G. (2003), Proton transport through water-filled carbon nanotubes, *Phys Rev Lett*, **90**, 105902-1–105902-4.
6. Bruni F., Ricci M.A., and Soper A.K. (1998), Water confined in vycor glass. I. A neutron diffraction study, *J Chem Phys*, **109**, 1478–1485.
7. Farrer R.A. and Fourkas J.T. (2003), Orientational dynamics of liquids confined in nanoporous sol-gel glasses studied by optical kerr effect spectroscopy, *Accounts Chem Res*, **36**, 605–612.
8. Cowan M.L., Bruner B.D., Huse N., Dwyer J.R., Chugh B., Nibbering E.T.J., Elsaesser T., and Miller R.J.D. (2005), Ultrafast memory loss and energy redistribution in the hydrogen bond network of liquid H_2O, *Nature*, **434**, 199–202.
9. Hibara A., Saito T., Kim H.-B., Tokeshi M., Ooi T., Nakao M., and Kitamori T. (2002), Nanochannels on a fused-silica microchip and liquid properties investigation by time-resolved fluorescence measurements, *Anal Chem*, **74**, 6170–6176.
10. Tas N.R., Haneveld J., Jansen H.V., Elwenspoek M., and van den Berg A. (2004), Capillary filling speed of water in nanochannels, *Appl Phys Lett*, **85**, 3274–3276.

11. Han A., Mondin G., Hegelbach N.G., de Rooij N.F., and Staufer U. (2006), Filling kinetics of liquids in nanochannels as narrow as 27 nm by capillary force, *J Colloid Interf Sci*, **293**, 151–157.
12. Haneveld J., Tas N.R., Brunets N., Jansen H.V., and Elwenspoek M. (2008), Capillary filling of sub-10 nm nanochannels, *J Appl Phys*, **104**, 014309-1–014309-6.
13. Whitby M., Cagnon L., Thanou M., and Quirke N. (2008), Enhanced fluid flow through nanoscale carbon pipes, *Nano Lett*, **8**, 2632–2637.
14. Chen X., Cao G., Han A., Punyamurtula V.K., Liu L., Culligan P.J., Kim T., and Qiao Y. (2008), Nanoscale fluid transport: size and rate effects, *Nano Lett*, **8**, 2988–2992.
15. Hibara A., Tsukahara T., and Kitamori T. (2009), Integrated fluidic systems on a nanometer scale and the study on behavior of liquids in small confinement, *J Chromatogr A*, **1216**, 673–683.
16. Kaji N., Ogawa R., Oki A., Horiike Y., Tokeshi M., and Baba Y. (2006), Study of water properties in nanospace, *Anal Bioanal Chem*, **386**, 759–764.
17. Schoch R.B. and Renaud P. (2005), Ion transport through nanoslits dominated by the effective surface charge, *Appl Phys Lett*, **86**, 253111-1–253111-3.
18. Schoch R.B., van Lintel H., and Renaud P. (2005), Effect of the surface charge on ion transport through nanoslits, *Phys Fluid*, **17**, 100604-1–100604-4.
19. Tsukahara T., Kuwahata T., Hibara A., Kim H.-B., Mawatari K., and Kitamori T. (2009), Electrochemical studies on liquid properties in extended nanospaces using mercury microelectrodes, *Electrophoresis*, **30**, 1–7.
20. Liu S., Pu Q., Gao L., Korzeniewski C., and Matzke C. (2005), From nanochannel-induced proton conduction enhancement to a nanochannel-based fuel cell, *Nano Lett*, **5**, 1389–1393.
21. van der Heyden F.H.J., Stein D., Besteman K., Lemay S.G., and Dekker C. (2006), Charge inversion at high ionic strength studied by streaming currents, *Phys Rev Lett*, **96**, 224502-1–224502-4.
22. van der Heyden F.H.J., Stein D., and Dekker C. (2005), Streaming currents in a single nanofluidic channel, *Phys Rev Lett*, **95**, 116104-1–116104-4.

23. van der Heyden F.H.J., Bonthuis D.J., Stein D., Meyer C., and Dekker C. (2006), Electrokinetic energy conversion efficiency in nanofluidic channels, *Nano Lett*, **6**, 2232–2237.
24. van der Heyden F.H.J., Bonthuis D.J., Stein D., Meyer C., and Dekker C. (2007), Power generation by pressure-driven transport of ions in nanofluidic channels, *Nano Lett*, 7, 1022–1025.
25. Morikawa K., Mawatari K., Kato M., Tsukahara T., and Kitamori T. (2010), Streaming potential/current measurement system for investigation of liquids confined in extended-nano space, *Lab Chip*, **10**, 871–875.
26. Kato M., Inaba M., Tsukahara T., Mawatari K., Hibara A., and Kitamori T. (2010), Femto liquid chromatography with attoliter sample separation in the extended nanospace channel, *Anal Chem*, **82**, 543–547.
27. Pu Q., Yun J., Temkin H., and Liu S. (2004), Ion-enrichment and ion-depletion effect of nanochannel structures, *Nano Lett*, **4**, 1099–1103.
28. Tsukahara T., Maeda T., Mawatari K., Hibara A., and Kitamori T. (2008), Study on vapor-liquid phase transition phenomena in extended-nano spaces, *Proc microTAS 2008*, 1311–1313.
29. Hibara A., Toshin K., Tsukahara T., Mawatari K., and Kitamori T. (2008), Microfluidic distillation utilizing micro-nano combined structure, *Chem Lett*, **37**, 1064–1065.
30. Tsukahara T., Mizutani W., Mawatari K., and Kitamori T. (2009), NMR studies of structure and dynamics of liquid molecules confined in extended nanospaces, *J Phys Chem B*, **113**, 10808–10816.
31. Tsukahara T., Hibara A., Ikeda Y., and Kitamori T. (2007), NMR study of water molecules confined in extended-nano spaces, *Angew Chem Int Edit*, **46**, 1180–1183.
32. Andoh Y., Kurahashi K., Sakuma H., Yasuoka K., and Kurihara K. (2007), Anisotropic molecular clustering in liquid ethanol induced by a charged fully hydroxylated silicon dioxide (SiO_2) surface, *Chem Phys Lett*, **448**, 253–257.
33. Tsukahara T., Mawatari K., Hibara A., and Kitamori T. (2008), Development of a pressure-driven nanofluidic control system and its application to an enzymatic reaction, *Anal Bioanal Chem*, **391**, 2745–2752.

34. Tsukahara T., Nagaoka K., and Kitamori T. (2008), Deuterium substitution and solvent effects on reaction dynamics in extended-nano spaces on a chip, *Proc microTAS 2008*, 1537–1539.
35. Malloggi F., Pannacci N., Attia R., Monti F., Mary P., Willaime H., Tabeling P., Cabane B., and Poncet P. (2009), Monodisperse colloids synthesized with nanofluidic technology, *Langmuir*, **26**, 2369–2373.
36. Huber D.E., Markel M.L., Pennathur S., and Patel K.D. (2009), Oligonucleotide hybridization and free-solution electrokinetic separation in a nanofluidic device, *Lab Chip*, **9**, 2933–2940.
37. Zevenbergen M.A.G. and Wolfrum B.L. (2009), Fast electron-transfer kinetics probed in nanofluidic channels, *J Am Chem Soc*, **131**, 11471–11477.
38. Emon H., Mawatari K., Tsukahara T., and Kitamori T. (2009), Surface modification of lipid bilayers in extended-nano space for making artificial intercellular structures, *Proc microTAS 2009*, 1524–1526.

Chapter 7

APPLICATION TO CHEMISTRY AND BIOTECHNOLOGY

7.1. Separation

7.1.1. *Separation by electrophoresis*

Electrophoresis is frequently used for DNA, or protein, separations. Due to the very easy fluidic control through simply applying voltage, it is also applied for extended nanospace. In conventional electrophoresis on microchips, target molecules were separated depending on their electrophoretic speed v, which is described by

$$v = \mu \frac{V}{L} = \mu E, \tag{1}$$

where μ is electrophoretic mobility, V is applied voltage, L is separation length, E is electric field, q is electrical charge, η is viscosity, and r is the hydrodynamic radius of the target molecule. In extended nanospace, the basic principle can be applied with characteristics of very small sample volume. In addition, several unique characteristics of the extended nanospace are utilized for separations.

Separation by nanopillars is one example used for large DNA separation.[1] Top-down fabrication methods by EB and DRIE were used to produce a very high aspect ratio and size-controlled nanopillars (diameter 500 nm and height 2700 nm) with 500 nm spacing on fused-silica substrates. The details of the microchip are shown in Figure 7.1. The spacing was controlled so as to be larger than the gyration radius, R_g, of DNA, and the nanopillars worked as the DNA serving matrix. By applying a DC electric field across

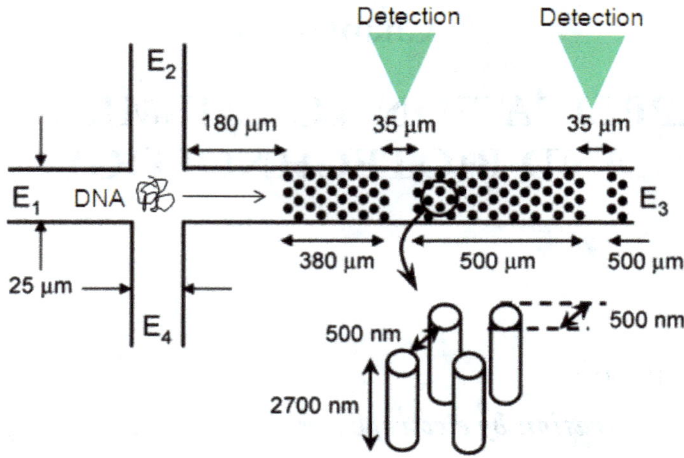

Figure 7.1. Nanopillar chips for long DNA separation (Adapted from Ref. 1.).

the nanopillars, DNA was introduced into the nanopillar area. Longer DNA strands migrated slower than shorter DNA strands due to the sieving effect of the nanopillars. In conventional gel electrophoresis, the maximum length of the DNA separation was limited to ~40 kbp, due to the difficulties of controlling the pore sizes of the gel in a wide range. However, the nanopillars can be fabricated and controlled by top-down technologies and allowed for separation of T4-DNA (165.6 kbp, $R_g = 970$ nm) and λ-DNA (48.5 kbp, R_g = 520 nm) due to the comparable size of R_g with the nanopillar spacing (500 nm). In addition, the separation time was reduced to just several seconds.

Interfaces of micro- and extended nanospace channels can also make new separation schemes, such as Ogston sieving, reptation, and entropic recoil, as shown in Figure 7.2.[2] Deep and shallow 1-D extended nanospace channels are arranged periodically, and the applied electric field acts as a driving force to move the molecules. For smaller molecules, the radius of gyration can be smaller than the channel dimensions. In this case, Ogston sieving becomes the dominant separation mechanism in which smaller molecules migrate faster than the larger molecules. If the channel dimensions are larger

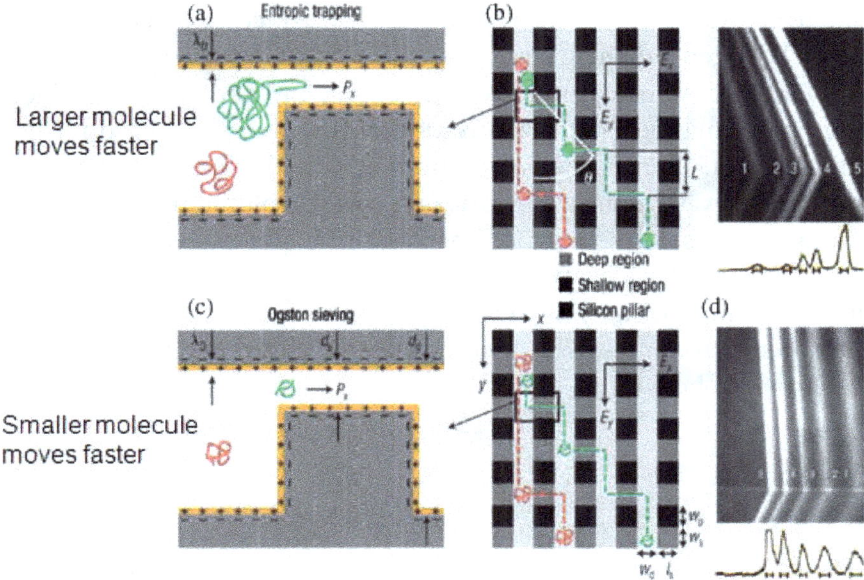

Figure 7.2. Separation of entropy trapping and Ogston seiving.

than the target molecules, entropic trapping will be dominant, since the possibility of entering, or escaping from, the 1-D extended nanospace channel dominates the separation process, allowing the larger molecules to migrate faster than the smaller molecules. The performance of this device was demonstrated by the variety of separations performed, including entropic trapping separation of macromolecules (λ-DNA-Hind III digest) and Ogston sieving of smaller PCR markers.

7.1.2. *Separation by pressure-driven flow or shear-driven flow*

Electrophoretic separations allow for easy handling of molecules and do not suffer from the large pressure drop when the channel dimensions are reduced, though the target is usually limited to positively or negatively charged ions in an aqueous solution. In contrast, pressure-driven flow (or shear-driven flow) can target a wider range of molecules in organic or aqueous phases and be applied to general chemistry.

Liquid chromatography is a separation method which is widely used in chemistry, while pressure-driven flow is often utilized for fluidic control. Previously, integration of liquid chromatography in microchips was attempted by utilizing packed column microchannels, though the resolution was not improved.[3] Height equivalent to a theoretical plate H is a parameter to show the separation performance of liquid chromatography, with lower values representing higher separation performance. HETP can be expressed by the van Deemter equation, as below:

$$H = \frac{L}{N} = A + \frac{B}{u} + Cu, \tag{2}$$

where N is the number of theoretical plates used to show the separation efficiency, L is the length of the separation column, A, B, and C are factors determined by the separation column, and u is the averaged linear velocity of a mobile phase. In conventional liquid chromatography, packed columns are used for achieving high surface-to-volume ratios, and all the A, B, and C terms affect H. However, the effect of the A and C terms can be reduced by reducing the channel size to a few hundred nanometers, as shown in Figure 7.3. This removes the limitations of the current liquid chromatography systems, namely the generation of heterogeneous pathways for the solute (eddy diffusion), and the resultant peak broadening by utilizing the channel surface as stationary phase. In addition to the improvement in the separation efficiency, unique separation mechanisms can be expected. The size of the extended nanospace becomes comparable to the length of the electric double layer, and the ions unevenly distribute inside the channel, depending on the charge of the molecules. Flow velocity in the center of the cross section is usually greater than that near the wall. Therefore, the counter-ions will move faster than the co-ions. By incorporating the separation principle, open column capillary chromatography was proposed.[4] For this, a fused-silica capillary column with a radius of several hundred nm was used.

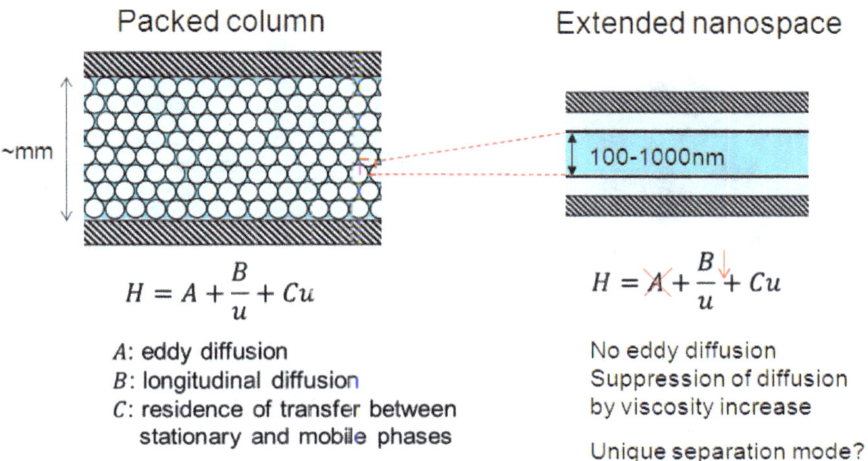

Figure 7.3. Improvement of separation performance by integrating liquid chromatography in extended nanospace.

Separation of fluorescent molecules was successfully demonstrated without applying an external voltage dependent on the charge. The excellent separation efficiency was demonstrated by approximately 100,000 plates per meter. Kitamori *et al.* reported integrated chromatography systems by utilizing 2-D extended nanospace on a microchip.[5] In this injection operations of aL samples and separation functions were integrated on a glass substrate (Figure 7.4). Extended nanospace channels were connected to four microchannels, to which air pressure was applied by a compressor and air-pressure controller. Fluorescent molecules were successfully separated depending on the charge, which was consistent with considering the surface negative charge. The efficient separation was shown in approximately 170,000 plates per meter. By eliminating the band-broadening in injection, more than one million plates per meter can be expected.

Shear-driven flow has also been applied for liquid chromatography in 1-D extended nanospace channels.[6] After the sample solution is introduced into the nm-sized gap between the upper and lower substrates using capillary action, the upper substrate is moved in the direction parallel to the lower plate. Materials such as a flat plate,

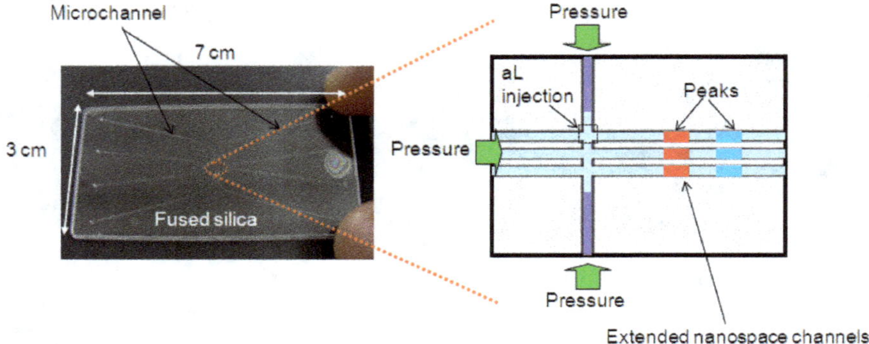

Figure 7.4. Injection of fL~aL sample and chromatography in a 2-D extended nanospace channel.

rotation disk, or flexible belt can be utilized as the moving upper substrates. Since the sample solution is automatically dragged in the gap according to the viscous effects of solution, a hydrodynamic flow, called a shear-driven flow, can be induced. The advantage of fluidic control by shear-driven flow is the relatively very fast flow velocity (~mms^{-1}) compared with the typical flow velocity (< mms^{-1}) of pressure-driven flow, which suffers from the large pressure drop. In this fluidic control mode, reversed-phase separation is better for preventing the leakage of mobile-phase aqueous solutions, and the surface of the extended nanospace channel is coated with a C18 monolayer. The number of theoretical plates for the separation of three Coumarin dyes was high (17900–24100), and very short separation time (< 1 s) was demonstrated, while conventional HPLC would require several minutes.[7]

7.2. Ion Transport

Control of ion transport is an important operation for analytical technologies. In microspace, ion transport is usually controlled by external forces, such as the electric field. However, when the size of the space is decreased to an extended nanospace, ion transport is largely affected by the surface wall charge of the channels, because the size becomes comparable to the length of the electric double layer. Therefore, several unique unit operations can be expected from utilizing the surface charge. For example, glass surfaces have negative charge in approximately neutral pH conditions, cations are enriched near the negatively charged glass surface, while anions are repelled and localized to the center of the nanochannel. In some conditions, anions are excluded from the nanochannels due to the large electrostatic repulsive forces. As an example, ion enrichment/depletion in 1-D nanochannels was demonstrated using the transport phenomena of fluorescent dye molecules.[8] By using these unique characteristics, various chemical operations can be realized, as described below.

Ion rectification is a unique unit operation which is quite difficult in microspace. Similar to diodes in electronic devices, ions in solutions can be rectified based on several principles. A direct approach is fabricating a gate electrode beside the extended nanospace channels, where large ion enrichment (e.g. enrichment of positively charged ions by the negatively charged channel wall) was induced and the direction of ion movement was controlled by an applied voltage between the extended nanospace channels.[9] The basic principle is quite similar to that of metal-oxide-semiconductor field-effect transistors (MOSFETs). By applying positive/negative voltage on the gate electrode, movement of the enriched positive/negative ions is restricted, which leads to ion rectification. Other unique approaches are proposed without the complicated fabrication processes of the gate electrode, as shown in Figures 7.5 and 7.6. In Figure 7.5, conical-shaped extended nanospace channels are fabricated between two microchannels. The surface negative charge affects the potentials of the ions. Due to the opening structure at the left side entrance, a large electric field is induced at the left side entrance, while a small electric field is induced at the right side entrance. Therefore, positive ions easily move to the right, while negative ions feel a large potential barrier when they move in the same direction. Based on this principle, ion rectification can be realized. This principle allows ion rectification even for pressure-driven flows. The other principle is based on partial surface modification and control of the applied voltage, as shown in

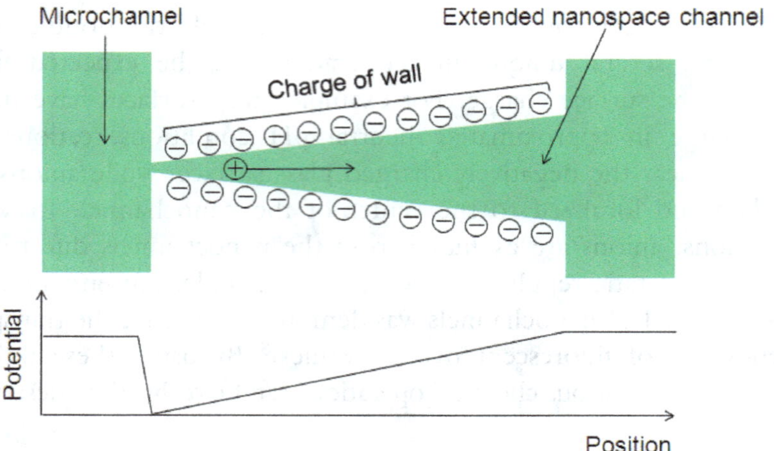

Figure 7.5. Principle of ion rectification by conical-shaped extended nanospace channel. The potential is shown for positive ions.

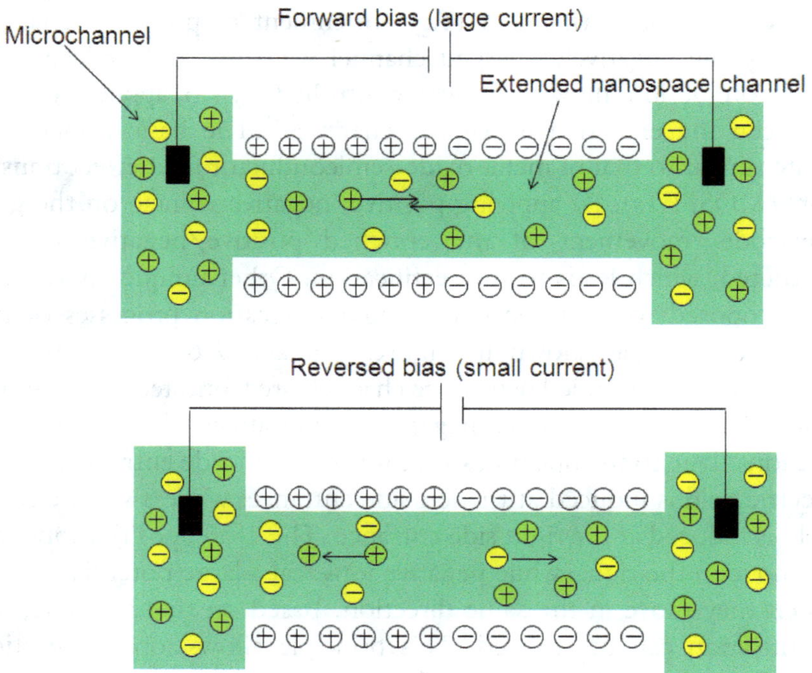

Figure 7.6. Principle of ion rectification by surface modification and applied voltage.

Figure 7.6.[10] By local surface modifications, the wall of the extended nanospace channel is locally positively charged on the left side, while it is negatively charged on the right side (similar to a p-n junction in diodes of electronic devices). Thus, negative and positive ions are enriched at the left and right sides, respectively. When a forwarded bias is applied between the extended nanospace channel, both negative and positive ions can pass through the extended nanospace channel and generate a large current. However, when a reverse bias is applied, a barrier layer is formed at the center of the extended nanospace channels, with neither positive nor negative ions able to pass the barrier. As such, the ion current becomes small. Based on this principle, ion rectification is realized. Additionally, there is also a combination of conical-shaped structures and local surface modification.

7.3. Concentration

Concentration is an operation which is usually used for sensitive analysis. For extended nanospace in particular, concentration is important due to the small absolute number of molecules resulting from the small volume. Several types of concentration are possible using extended nanospace. The first is solid phase extraction. In conventional solid phase extraction, microbeads are packed in a column, and a large amount of sample is introduced to concentrate the target molecules on the microbead surfaces (Figure 7.7). The large surface-to-volume ratio of the packed columns allows concentration of the target molecules in a very small space. This strategy for concentration can be further applied to the extended nanospace. Smaller particles are packed in the microchannels, and extended nanospace channels can work as a filter with very small size. For example, 60 nm gold nanoparticles were packed in a microchannel with an extended nanospace channel of diameter 40 nm.[11] β-amyloid peptides were concentrated on the nanoparticles and selectively detected by a surface-enhanced Raman scattering (SERS) with a high sensitivity (~nM), sufficient to analyze cerebrospinal fluid (CSF) and blood plasma. In another example, microtubles were utilized to selectively concentrate target molecules. The microtubles were functionalized with

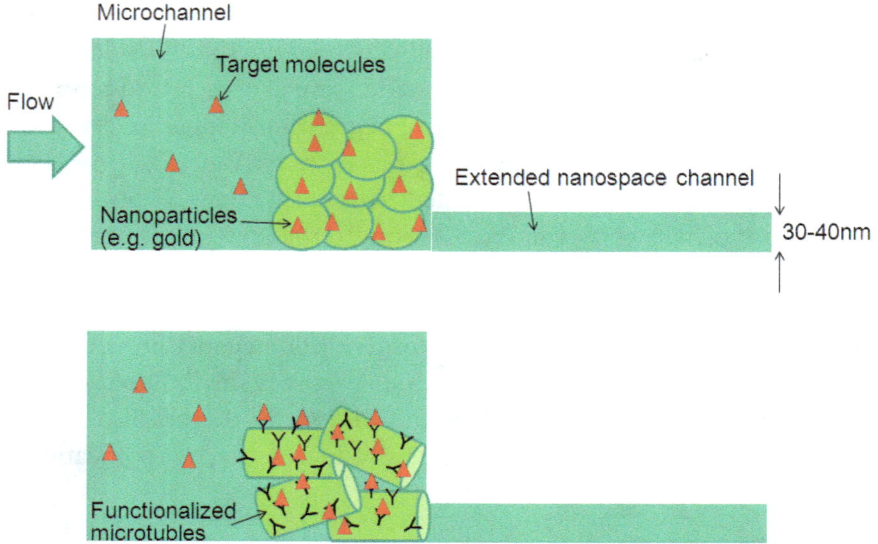

Figure 7.7. Principle of concentration of target molecules by an extended nanospace filter.

antibodies by chemical bonding and trapped at the entrance of the extended nanospace channel 30 nm deep. Proteins were selectively trapped by immunoreactions. A maximum of five orders of concentration were achieved.

Other unique concentration operations can be realized by utilizing the ion enrichment/depletion effect by the surface charge of the channels. A representative example is shown in Figure 7.8. When a voltage is applied across the extended nanospace channel with negatively charged walls, enrichment of positive ions occurs in the extended nanochannels, while negative ions are depleted. By increasing the applied voltage, the ion enrichment/depletion area expands to the microchannels adjacent to the exit of the extended nanospace channel. By applying the voltage between the microchannels, negatively charged target ions (FITC fluorescent dyes) can be concentrated at the interface in front of the ion depletion area, as shown in Figure 7.8. Almost one million times concentration was achieved, which is promising for the very sensitive analysis of target molecules.

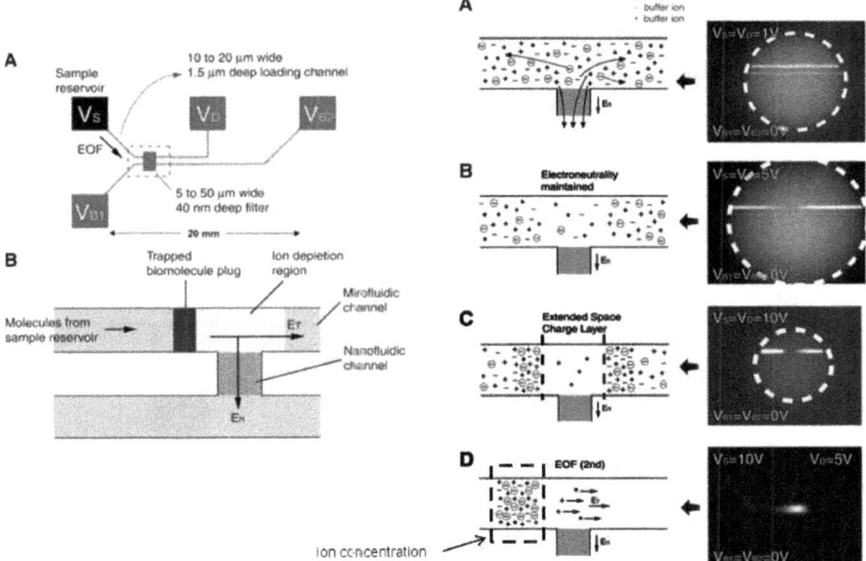

Figure 7.8. Principle of concentration by ion enrichment/depletion and EOF.

7.4. Single Molecule Handling

By utilizing the extremely small space volume of extended nanochannels, sensitive analysis of DNA molecules is possible. Nanofluidic systems have been used to probe conformational, dynamic, and entropic properties of DNA molecules, to rapidly sort DNA molecules based on length-dependent interactions with their confining environment, and for determining the spatial location of genetic information along long DNA molecules. In this section, recent experiments utilizing fluidic systems comprised of nanochannels, nanoslits, nanopores, and zero-mode waveguides for DNA analysis are reviewed.[12]

7.4.1. *DNA separation*

Here, we introduce an extended nanofluidic channel device, consisting of many entropic traps, designed and fabricated for the separation of long DNA molecules.[13] The basic design of the entropic trap array

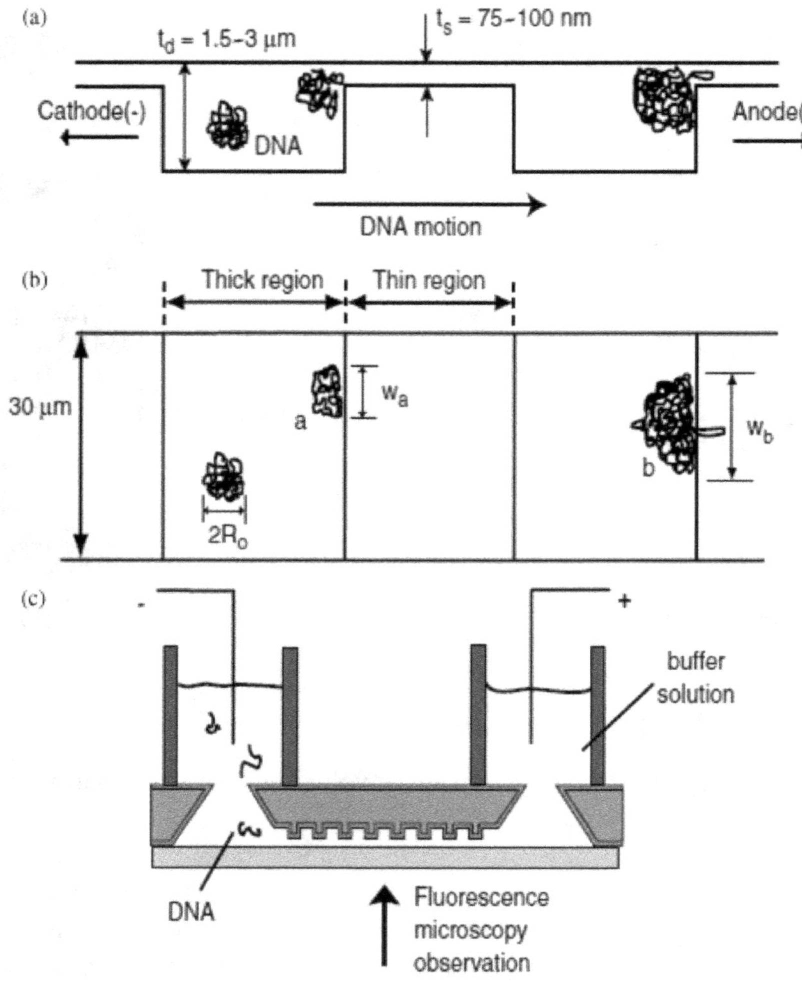

Figure 7.9. Nanofluidic separation device with many entropic traps: (a) Cross-sectional schematic diagram of the device, (b) Top view of the device in operation, (c) Experimental setup. Reservoirs are made at both ends of the channel and filled with DNA solution. (Redrawn from Ref. 13).

(Figure 7.9(a)) consists of alternating thin and thick regions in a microfabricated channel. The channel depth of the thin region is smaller than the radius of gyration, R_o, of the DNA molecules being separated, and thus it serves as a molecular sieve. In the thick region,

DNA molecules can form spherical equilibrium shapes as the thickness is larger than R_o, whereas in the thin region, DNA molecules are deformed. When driven by an electric field, the DNA molecules travel through alternating thick and thin regions, repeatedly changing their conformation. This conformation change costs entropic free energy, so DNA molecules are temporarily trapped at the entrance of the thin regions. This entropic trapping limits the overall mobility of DNA molecules in the channel, and the mobility of the DNA becomes length dependent. Interestingly, longer DNA molecules actually have higher mobility in this channel.[14] In the escape of a DNA molecule from an entropic trap, only the part of the molecule that is in contact with the boundary of the thin region plays a crucial role. Whenever a sufficient number of DNA monomers are introduced into the high-field thin region (by Brownian motion), the escape of the whole molecule is initiated.[15] Longer DNA molecules, with larger R_o, have a larger surface area in contact with the boundary and therefore have a higher probability to escape per unit time (due to a higher escape attempt frequency), which leads to a shorter trapping time and higher overall mobility (Figure 7.9(b)). The channels are fabricated using photolithography and etching techniques on a Si substrate. Fluorescently labeled DNA solutions are loaded into the reservoirs, and then introduced into the channel by electrophoresis (Figure 7.9(c)). The motion of individual DNA molecules, as well as DNA bands, is observed with optical microscopy. This process creates electrophoretic mobility differences, thus enabling efficient separation without the use of a gel matrix or pulsed electric fields. Samples of long DNA molecules (5000 ~160,000 base pairs) were efficiently separated into bands in 15 mm long channels. Multiple-channel devices operating in parallel were demonstrated. The efficiency, compactness, and ease of fabrication of the device suggest the possibility of more practical integrated DNA analysis systems.

7.4.2. *Unfolding of DNA*

Next, we introduce the entropic unfolding of DNA molecules in extended nanofluidic channels.[16] Single DNA molecules confined to extended nanofluidic channels extend along the channel axis in order to minimize their conformational free energy. When such molecules are forced into an extended nanochannel under the application of an external electric field, monomers near the middle of the DNA molecule may enter first, resulting in a folded configuration with less entropy than the unfolded molecule. In this report, DNA molecules were inserted and observed folded in nanochannels, as depicted in Figure 7.10. Long DNA molecules are electrophoretically driven from a microfluidic region to the entrance of the smaller channel. The electric field is turned off before the molecules enter the channel and thermal agitation causes them to undergo a number of different

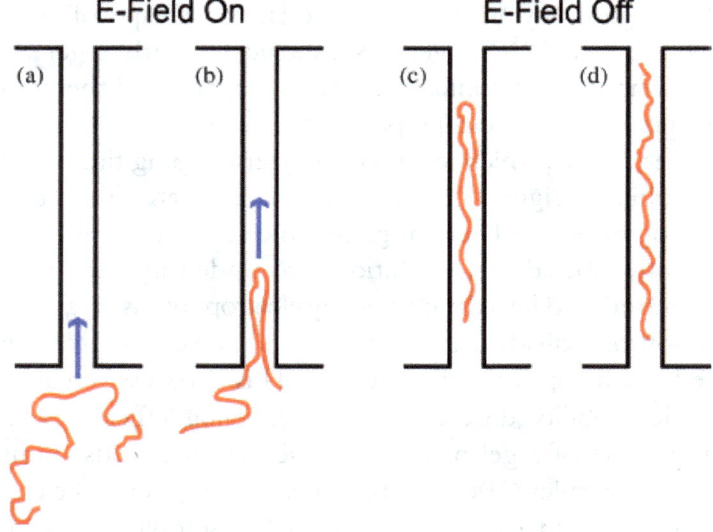

Figure 7.10. Overview of experimental procedure: (a) A long DNA molecule sits in a microchannel, adjacent to the entrance of a nanochannel, (b) The electric field (blue arrow) pulls the DNA into the nanochannel. Because the molecule's entrance was initiated at some point along the backbone distinct from either of the ends, it enters the channel in a folded conformation, (c) Once the electric field is turned off, the DNA strand relaxes inside the nanochannel in a high energy folded state. It gradually unfolds, thereby reducing its conformational free energy, (d) The molecule has completely unfolded within the channel, extending to its equilibrium conformation (Redrawn from Ref. 16).

intramolecular conformations. When a molecule happens to be in an appropriate conformation and position relative to the channel entrance (as depicted in Figure 7.10(a)), the field is turned on and the molecule enters the channel, often with a folded front end. After the entire DNA molecule has entered the channel, the field is switched off and the dynamics of the molecule are observed. The increased free energy of a folded molecule results in two effects: an increase in the extension factor per unit length for each segment of the molecule, and a spatially localized force that causes the molecule to spontaneously unfold. The ratio of this unfolding force to the hydrodynamic friction per DNA contour length is measured in nanochannels with two different diameters.

7.5. Bioanalysis

Since the volume of extended nanospace is extremely small compared with conventional devices, even smaller than single cells, this special space provides a kind of ideal analysis space. In this section, we introduce some bioanalyis methods exploiting this property of extended nanospace.

7.5.1. *Immunoassay*

Here, we introduce a novel ELISA (enzyme-linked immuno sorbent assay) system utilizing extended nanospace (3 μm wide and 300 nm deep) for analysis at the single molecule level.[17] A new method for understanding gene and protein expression at single cell level is increasingly desired in the fields of proteomics, metabolomics research, and diagnostics. One of the challenges is a lossless, widely applicable molecular recognition with single-molecule sensitivity and a large number of co-existing molecules. One of the most selective molecular recognition methods is immunoassay. ELISA, especially, is frequently utilized due to the high sensitivity from the amplification reaction of dye molecules with the enzyme. Previously, Kitamori and coworkers reported the integration of the ELISA system on a microchip and verified the fast and sensitive analytical performance, combining a thermal lens microscope as an ultra sensitive detector for non-fluorescent molecules.[18] In this report, they developed a format for repeatable immunoassay in extended nanospace, and investigated the performance. Approximately 20 molecules (in a 60 pL sample volume) were detected.

Micro- and nanochannel design was fabricated on a glass substrate, as shown in Figure 7.11. The sample and reagents were introduced into the microchannel through pressure (100 kPa) and introduced into the nanochannel by further increasing the pressure (400 kPa). In the extended nanospace channel, ELISA was realized. For repetitive immunoassays, capture antibodies were immobilized by

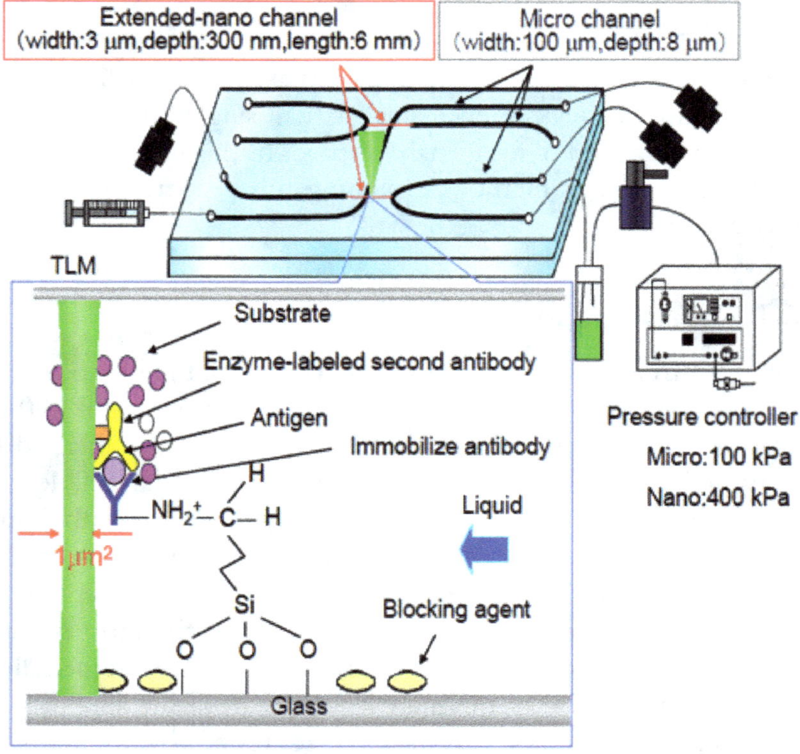

Figure 7.11. Schematic illustration of a novel ELISA system inside an extended nanospace channel.

chemisorptions, and glycine-HCl (pH = 1.5) was introduced to remove the antigen. In this way, immunoassays (changing the antigen concentration) were conducted repeatedly.[19] For detection, substrates were introduced and dye molecules produced by enzymatic reaction were detected by TLM. The sample volume introduced was controlled through measuring the flow velocity in the nanochannel by introducing fluorescent solutions and measuring the intensity at two points (300 μm distance apart) in the nanochannel. The performance was evaluated, and a limit of detection (LOD) of 20 molecules (in a 60 pL sample volume) was determined. This system will become a powerful tool for single cell analysis due to the relatively small volume.

7.5.2. *DNA analysis*

Single DNA detection has become a focus in genomics, diagnosis, and medical genetics. Information of each single molecule in each single cell is necessary in medical genetic diagnoses. Rolling circle amplification (RCA) is one of the most popular methods to detect single DNA molecules,[20] however, it is difficult to apply to ultra small volume samples, such as single cells. Here, we introduce a patterning and detection method for single molecule DNA in extended nanochannels for small volume sample analysis.[21] RCA protocol is shown in Figure 7.12. The target DNA was recognized after linear padlock probe hybridization with the probe ends joined through ligation to their target sequences. At first, the primer was immobilized to the surface of the modified glass substrate, then, linear padlock probes were hybridized to the immobilized primer. The padlock was designed for the detection of target DNA, and could be hybridized to all target DNA sequences. The DNA circle was produced by ligation on varying concentrations of the target DNA, followed by RCA using Ampligase. Finally, to label the amplified DNA molecules, fluorescent probe DNA was introduced into the microchannel and hybridized.

As a result, the RCA product was visualized in extended nanochannels. The number of fluorescent dots changed dependent on the target DNA concentration. In this report, RCA in extended nano closed space was demonstrated for the first time.

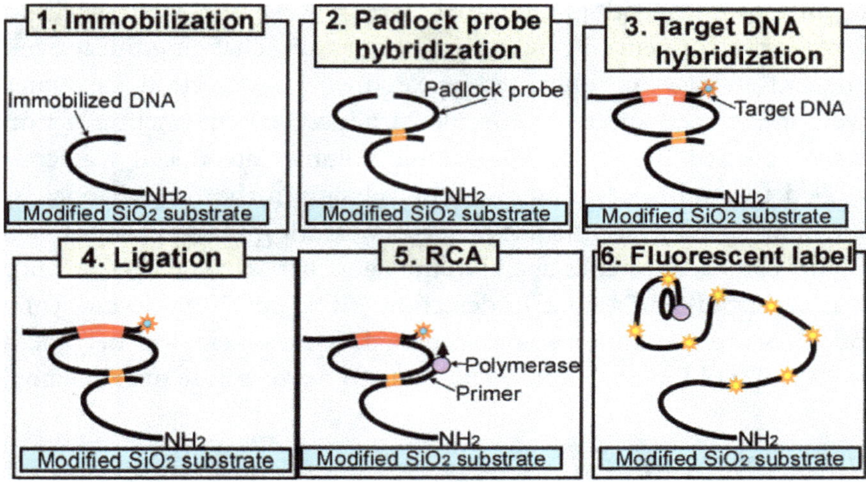

Figure 7.12. Scheme of RCA on microchannel glass surface: 1. Immobilization of primer DNA on glass surface, 2. Padlock probe introduction and hybridization to an immobilized primer DNA, 3. Target DNA introduction and hybridization to the padlock probe, 4. Ligation of the padlock probes, 5. RCA using polymerase, 6. Fluorescent labeling by hybridization of fluorescently labeled DNA (Redrawn from Ref. 21).

7.6. Energy Devices

As described in the previous chapters, the size of the extended nanospace can become comparable with that of the electric double layer and co-ions and counter-ions are then unevenly distributed, depending on the charge of the wall. When the counter-ions in the electric double layer are transported into the extended nanospaces by the pressure-driven hydrodynamic flow, it induces a streaming potential/current. This means that the fluidic energy can be easily and automatically converted to electric power. Energy devices using this principle will become novel hydraulic power generation devices, with a very simple structure compared with conventional macroscale plants. The power generation was actually demonstrated by utilizing 1-D extended nanospace channels on fused-silica substrate.[22] By adjusting the ion strength and load resister a power of 240 pW was obtained for a single channel (width: 50 μm, depth: 490 nm and length: 4.5 mm) by applying a pressure of 3 bars. A peak efficiency of 3 per cent was obtained for the 70 nm deep channel, when the electric double layer was overlapped.

The extended nanospace can also be applied to fuel cell devices. Matzke *et al.* reported that the apparent proton conductivity inside a nanochannel was enhanced by several orders of magnitude due to the electric double layer overlap utilizing 1-D extended nanochannels.[23] This unique property was used for fuel cell applications with the extended nanochannel working as a conventional proton exchange membrane. In total, 55 parallel extended nanospace channels were arrayed to bridge the two U-type microchannels. The depth of the extended nanochannels was changed from 50 nm to 50 μm, with a constant width of 100 μm. The apparent conductivity of the protons increased below 1 μm, while the size for the transition was larger than the length of the electric double layer (approximately 100 nm) calculated by conventional Debye–Huckel theory. The reason for this increase is not clear, however. Kitamori *et al.* suggested that size confinement of water enhanced the proton exchange rate, which was 20 times larger than in bulk water. This increase was also observed for 800 nm,[24] and they introduced a proton exchange layer and proposed a three-layer model to explain the size effect. These phenomena might be closely related, and more detailed study is required. These nanofluidic devices could be applied to fuel cell applications by densely packing the extended nanochannels working as proton exchange membranes.

References

1. Kaji N., Tezuka Y., Takamura Y., Ueda M., Nishimoto T., Nakanishi H., Horiike Y., and Baba Y. (2004), Separation of long DNA molecules by quartz nanopillar chips under a direct current electric field, *Anal Chem*, 76, 15–22.
2. Fu J., Schoch R.B., Stevens A.L., Tannenbaum S., and Han J. (2007), A patterned anisotropic nanofluidic sieving structure for continuous-flow separation of DNA and proteins, *Nat Nanotechnol*, **2**, 121–128.
3. Ericson C., Holm J., Ericson T., and Hjerten S. (2000), Electroosmosis- and pressure-driven chromatography in chips using continuous beds, *Anal Chem*, **72**, 81–87.

4. Wang X., Kang J., Wang S., Lu J.J., and Liu S. (2008), Chromatographic separations in a nanocapillary under pressure-driven conditions, *J Chromatogr A*, **1200**, 108–113.
5. Kato M., Inaba M., Tsukahara T., Mawatari K., and Kitamori T. (2010), Femto liquid chromatography with attoliter sample separation in the extended nanospace channel, *Anal Chem*, **82**, 543–547.
6. Vankrunkelsven S., Clicq D., Cabooter D., Malsche W.D., Gardeniers J.G.E., and Desmet G. (2006), Ultra-rapid separation of an angiotensin mixture in nanochannels using shear-driven chromatography, *J Chromatogr A*, **1102**, 96–103.
7. Fekete V., Clicq D., Malsche W.D., Gardeniers H., and Desmet G. (2008), Use of 120 nm deep channels for liquid chromatographic separations, *J Chromatogr A*, **1189**, 2–9.
8. Pu Q., Yun J.S., Temkin H., and Liu S.R. (2004), Ion-enrichment and ion-depletion effect of nanochannel structures, *Nano Lett*, **4**, 1099–1103.
9. Daiguji H., Yang P., and Majumdar A. (2004), Ion transport in nanofluidic channels, *Nano Lett*, **4**, 137–142.
10. Daiguji H., Oka Y., and Shirono K. (2005), Nanofluidic diode and bipolar transistor, *Nano Lett*, **5**, 2274–2280.
11. Chou I.H., Benford M., Beier H.T., Cote G.L., Wan M., Jing N., Kameoka J., and Good T.A. (2008), Nanofluidic biosensing for β-amyloid detection using surface enhanced raman spectroscopy, *Nano Lett*, **8**, 1729–1735.
12. Levy S.L. and Craighead H.G. (2010), DNA manipulation, sorting, and mapping in nanofluidic systems, *Chem Soc Rev*, **39**, 1133–1152.
13. Han J. and Craighead H.G. (2000), Separation of long DNA molecules in a microfabricated entropic trap array, *Science*, **288**, 1026–1029.
14. Han J. and Craighead H.G. (1999), Entropic trapping and sieving of long DNA molecules in a nanofluidic channel, *J Vac Sci Technol A*, **17**, 2142–2147.
15. Han J., Turner S.W., and Craighead H.G. (1999), Entropic trapping and escape of long DNA molecules at submicron size constriction, *Phys Rev Lett*, **83**, 1688–1691.
16. Levy S.L., Mannion J.T., Cheng J., Reccius C.H., and Craighead H.G. (2008), Entropic unfolding of DNA molecules in nanofluidic channels, *Nano Lett*, **8**, 3839–3844.

17. Hiruma F., Mawatari K., Tsukahara T., and Kitamori T. (2008), Integration of immunoassay into extended nanospace for analysis at single-molecule level, *Proc microTAS 2008*, 221–223.
18. Sato K., Yamanaka M., Hagino T., Tokeshi M., Kimura H., and Kitamori T. (2004), Microchip-based enzyme-linked immunosorbent assay (microELISA) system with thermal lens detection, *Lab Chip*, **4**, 570–575.
19. Zheng G., Patolsky F., Cui Y., Wang W.U., and Lieber C.M. (2005), Multiplexed electrical detection of cancer markers with nanowire sensor arrays, *Nat Biotechnol*, **23**, 1294–1301.
20. Jarvius J., Melin J., Göransson J., Stenberg J., Fredriksson S., Gonzalez-Rey C., Bertilsson S., and Nilsson M. (2006), Digital quantification using amplified single-molecule detection, *Nat Methods*, **3**, 725–727.
21. Tanaka Y., Xi H., Sato K., Mawatari K., Renberg B., Nilsson M., and Kitamori T. (2010), Extended-nano channel based rolling circle amplification to detect single molecule DNA, *Proc microTAS 2010*, 1160–1162.
22. van der Heyden F.H.J., Bonthuis D.J., Stein D., Meyer C., and Dekker C. (2007), Power generation by pressure-driven transport of ions in nanofluidic channels, *Nano Lett*, 7, 1022–1025.
23. Liu S., Pu Q., Gao L., Korzeniewski C., and Matzke C. (2005), From nanochannel-induced proton conduction enhancement to a nanochannel-based fuel cell, *Nano Lett*, **5**, 1389–1393.
24. Tsukahara T., Mizutani W., Mawatari K., and Kitamori T. (2009), NMR studies of structure and dynamics of liquid molecules confined in extended nanospaces, *J Phys Chem B*, **113**, 10808–10816.

Chapter 8

FUTURE PROSPECTS

Many researchers have tried to miniaturize chemical operations, and important chemical techniques such as analysis, diagnosis, and synthesis have been integrated into microfluidic chips. These microfluidic devices have demonstrated superior performance with shorter processing times (reduced from days or hours to minutes or seconds), smaller sample volumes (from mL scale to less than μL scale), and easier operation (from professional to personal) compared to conventional systems. However, the main advantage of such microfluidic devices is simply the large surface-to-volume ratio of the microspaces, and that the physicochemical properties of liquids in microspaces do not differ from those in the bulk: the operating principle is dominated by classical mechanics.

On the other hand, extended nanospace is believed to be a technologically and scientifically unexplored and attractive region, since this space is located between single nanometer technologies and microfluidic technologies, and is also a transition area of molecular behavior from an individual molecule to a bulk condensed phase. As mentioned in the former sections, new methodologies and technologies for fabrication, fluidic control, and detection for studying extended nanospaces have been established, and confinement-induced nanospatial properties, such as unique liquid and fluidic phenomena, differing from bulk, micro and nanospaces, have been found.

Since the performance of top-down nanofabrication techniques and surface modification techniques with nanometer scale has drastically improved, these techniques make it possible to create complicated and partially functional, patterned nanoscale channels and structures. When single nm-sized functional materials, such as

CNT, porous silica, proteins, and bio-polymers, are immobilized on top-down fabricated extended nanospace structures by using bottom-up methods, analysis, separation, and synthesis of single molecules under liquid condensed phase conditions are expected to be possible.

Unique liquid and fluidic properties are seen in extended nanospaces, but some aspects of nanofluidics and nanochemistry remain unexplored and obscure. In particular, since the higher viscosity and higher proton mobility of liquids inside extended nanospaces should be influenced by slipping and the electrokinetic phenomena of the fluids, the flow profile inside extended nanospaces will differ from the Hagen–Poiseuille flow observed in pressure-driven flows. Thus, new detection methods that can observe the flow profile with nanometer-scale spatial resolution are required. Conventional optical detection methods are unable to accurately measure the flow profile inside extended nanospaces, because of the general diffraction limit of about 1 μm, with the exception of some new approaches. Conventional PIV methods, in which nanoparticles are utilized for visualizing the flows, are also unable to correctly map the flow velocity profiles in extended nanospaces, due to the incomplete detection of flow profiles caused by large electrokinetic or entropic effects on the particles. Development of new methods without particles is essential to correctly observe the flow profiles in extended nanospaces.

These extended nanospace sciences and technologies are expected to create many innovative nanochemical and nanobiochemical devices. For example, the volume of 100 nm-sized extended nanospaces, which is on the order of attoliters aL = 10^{-18} L = $(100\ \text{nm})^3$ is far smaller than the volume of a single cell, which is in the order of picoliters (pL = 10^{-12} L). Furthermore, a solution with μM (10^{-6} M = molL^{-1}) order concentration occupying a 100 nm-sized extended nanospace contains molecular numbers approaching the single molecule level (10^{-24} mol). Moreover, large surface-to-volume ratios of extended nanospaces make it possible to capture individual molecules on functional ligands on the surface. By combining these characteristics, single cell immunoassays, where various kinds of biomolecules (such as proteins) produced from one cell can be handled and analyzed at single molecule level, will be realized in extended nanospaces.

We believe that research outcomes resulting from the study of extended nanospace will have important consequences for not only the establishment of new nanoscience, but also the establishment of new-concept nanotechnology devices.

INDEX

www.ingramcontent.com/pod-product-compliance
Lightning Source LLC
Chambersburg PA
CBHW050628190326
41458CB00008B/2189

* 9 7 8 1 8 4 8 1 6 8 0 1 5 *